Beginner's Guide to Artificial Intelligence (AI)

Understanding AI Essentials, Mastering AI Prompts, & Future-Proofing Your Career & Business

Roger Best

Disclaimer & Disclosure

The content provided in this book, including all examples,illustrations, prompts, exercises, and discussions, is intended foreducational, informational, and illustrative purposes only.
Neither the author nor any AI tools involved in the development of this material are licensed legal, financial, accounting, or regulatory professionals.

While every effort has been made to provide helpful, thoughtful, and practical guidance, the information contained herein should **not** be construed as legal advice, financial advice, investment advice, tax advice,or any other form of professional counsel.

Situations involving legal, financial, or regulatory decision-making are complex and highly individual. **You should always seek advice from qualified professionals** before making decisions that could impact your legal standing, financial health, business operations, or personal life.

The sample prompts, exercises, and workflows presented inthis book are intended as **reference materials and training illustrations only**. They are designed to spark creativity, expand understanding, and serve as starting points for your own exploration — **not as finalized solutions, definitive action steps, or comprehensive advice**.

Readers are solely responsible for their own actions ,decisions, and outcomes. The author and contributors disclaim any liability for any loss, injury, or damage that may result from the use or misuse of the content provided.

If you are unsure about how any of the material in this book might apply to your specific situation, or if you require advice on legal, financial, tax, regulatory, or business matters, **please consult a licensed and qualified professional** before proceeding.

By engaging with the material in this book, you cknowledge and agree to these terms.

Contents

Introduction

YOU'RE RUNNING LATE FOR a crucial meeting, frantically searching through endless files on your computer. Suddenly, your AI assistant chimes in, "Hey there! I noticed you're in a bit of a pickle. Want me to find that pesky file for you?" With a sigh of relief, you reply, "Yes, please!" In mere seconds, the file appears on your screen, and you're ready to conquer the day.

This might sound like a scene from a sci-fi movie, but it's not. AI is here, and it's changing the game for everyone, not just the tech gurus.

I remember my first brush with AI. It was like stumbling into a foreign land where everyone spoke a language I couldn't quite grasp. Terms like "machine learning" and "neural networks" flew over my head, leaving me feeling like a kid in a quantum physics class. But as I dug deeper, I realized that AI wasn't just some mysterious entity reserved for the tech elite. It was a tool, a companion, and a catalyst for change that could benefit anyone willing to learn.

That's why I wrote this book. I wanted to create a guide to demystify AI and show you, the everyday hero, how to wield its power for good. Whether you're an entrepreneur looking to streamline your business, a professional aiming to boost your productivity, or simply a curious soul eager to explore the frontiers of technology, this book is for you.

We'll dive into the nitty-gritty of AI, breaking down complex concepts into bite-sized pieces that are easy to digest. You'll learn how to master the art of prompt engineering, crafting commands that make AI work wonders for you. We'll explore real-world applications of AI, from automating tedious tasks to unlocking new revenue streams. And, of course, we'll tackle the big question on everyone's mind: how can AI safeguard your career and ensure you stay ahead of the game?

But this isn't just another dry, technical manual. We'll spice things up with hands-on exercises, real-world examples, and success stories that will make you feel like an AI superhero. You'll discover that learning AI isn't about memorizing jargon or writing complex code. It's about understanding the principles, asking the right questions, and applying the tools in a way that makes sense for you.

So, whether you're a complete novice or someone who's dabbled in AI but feels like you're still missing the big picture, buckle up. We're about to kick-start a journey that'll transform the way you think about technology and your own potential.

In the coming chapters, we'll explore the fascinating world of AI together. We'll start with the basics, unraveling the mystery behind terms like "algorithms" and "data." Then, we'll move on to the practical stuff, like using AI to automate email responses, generate content ideas, and even predict customer behavior. We'll also tackle some of the bigger questions, like the ethical implications of AI and how to ensure that it's used for good.

By the end of this book, you'll have a toolkit of AI techniques and a mindset that embraces innovation. You'll no longer feel intimidated by AI; instead, you'll feel empowered to leverage its potential in your own life and work.

So, are you ready to become an AI trailblazer? To step into a world where technology is not just a tool, but a partner in your success? Then let's dive in. The future is waiting, and it's brighter than you ever imagined.

Chapter One

Demystifying AI

Foundation & Essentials

EVER SLEPT THROUGH YOUR alarm and woken up to the scent of fresh coffee acting as your personal rescue mission? A good friend of mine told me this exact story—not about his alarm, mind you, but about how his AI-powered coffee maker picked up the slack. Without so much as a yawn, it sprang to life at 7 a.m., brewed his morning cup, and even cheekily recommended a double espresso—as if it knew what kind of day he was in for. Now, I won't claim AI will keep your life entirely on track (your socks are still your responsibility), but it can handle a surprising number of the little things that trip us up. And that's what we're getting into next: the curious, clever, and occasionally caffeinated world of artificial intelligence—and how it's quietly becoming your behind-the-scenes co-pilot.

What is AI? A Simple Breakdown for Beginners

So, what exactly is this mystical thing called AI? Let's break it down as if explaining it to your technophobic aunt. Imagine AI as a digital brain, tirelessly working to mimic human intelligence. While it doesn't have dreams of becoming a rock star or an existential crisis at three in the morning, it does excel at tasks like recognizing patterns and learning from experience. Think of it as that overachieving friend who always scores full marks, not because they know everything, but because they're really good at figuring things out.

AI's journey kicked off in 1956 at the Dartmouth Conference, where a bunch of brilliant minds got together and coined the term "artificial intelligence". This was the beginning of a tech revolution, setting the stage for a future where machines could think, or at least, give a pretty good impression of it. Fast forward to 1997, IBM's Deep Blue made history by defeating chess champion Garry Kasparov. Though Kasparov might have wished for a rematch, Deep Blue proved a point: machines were no longer just calculating numbers, they were strategizing like pros.

Now, let's talk about the different species of AI creatures. We've got Narrow AI, which is super focused on one task—like a cat laser-focused on its favorite toy. Siri and Alexa are examples, excelling at specific tasks but not much else. I find it particularly frustrating when I can't even tell Alexa to stop giving me suggestions about things it believes I might want to buy... but I think I may have digressed just a bit. Then, there's the theoretical Artificial General Intelligence (AGI), which aims to perform any intellectual task a human can. And if you want to go full sci-fi, there's Super AI, a concept where machines might surpass human intelligence altogether.

On the functionality front, we have Reactive Machines, like IBM's Deep Blue, which live in the moment and base decisions on current inputs without learning from the past. Limited Memory AI, on the other hand, can remember a thing or two, much like your goldfish recalling where the food is after you've fed it for the umpteenth time. Finally, there's the Theory of Mind AI and Self-Aware AI, both of which reside in the realm of sci-fi, rumored to understand emotions and possess self-awareness, but for now, they're as real as unicorns.

Today, AI is as common as coffee stains on a Monday shirt. From smartphones predicting your next word to smart home devices setting the perfect mood lighting, AI is everywhere, quietly revolutionizing how we live and work. It's not just about making things faster or more efficient; it's about enhancing our capabilities, offering new ways to think and solve problems. AI is no longer an optional upgrade; it's becoming an integral part of our tech-savvy lifestyles, powering a world where possibilities are as limitless as your imagination.

Exercise: AI Scavenger Hunt

Take a moment to look around your home or workplace. List three devices or applications using AI. Think beyond obvious gadgets like smartphones and consider software or services that might be leveraging AI behind the scenes. Reflect on how these tools impact your daily routine and consider how life might be different without them.

Key Components of AI: Algorithms, Neural Networks, and Beyond

Let's talk about algorithms. They're not as scary as they sound. Think of them as recipes for your favorite dish. Just like a recipe guides you through the steps to make Grandma's secret lasagna, an algorithm is a set of instructions for a computer to solve a problem. Whether it's sorting your emails by priority or helping a robot vacuum navigate your living room without sucking up your socks, algorithms are the unsung heroes. They make decisions based on logical steps, much like how you decide to eat dessert first because, well, life is short. Sorting algorithms organize data, while search algorithms help find information. They're the backbone of many AI applications, quietly orchestrating the complex symphony of data behind the scenes.

Now, onto neural networks, which sound like something out of a sci-fi flick. But they really are a clever bit of tech that mimics the human brain, albeit without the existential dread. Imagine your brain as a massive network of neurons, each firing off signals to process information. Neural networks work similarly, using layers of interconnected nodes to analyze and learn from data. They excel at recognizing patterns, like identifying your cat in a photo or distinguishing between a hot dog and a sandwich (a critical skill, indeed). Image recognition is a prime example of neural networks in action, as they sift through countless pixels to tell you if that's your Aunt Marge or a cleverly disguised raccoon.

Data, dear reader, is the fuel that powers the AI engine. It's like the flour in your bread or the beans in your chili—without it, nothing happens. Big data refers to those colossal amounts of information that traditional databases can't handle. AI thrives on big data, learning from patterns and trends to make predictions with impressive accuracy. But before AI can work its magic, data must be collected and preprocessed, cleaned up like a toddler after a spaghetti dinner. This ensures the information used is reliable and relevant, allowing AI to perform its tricks without tripping over messy inputs.

Machine learning models are the secret sauce that ties all these components together. They come in different flavors, mainly supervised and unsupervised learning. Supervised learning involves teaching the model with labeled data, like a teacher guiding a student with a textbook. Think of it as training a dog with treats—positive outcomes lead to rewards, reinforcing the behavior. Unsupervised learning, on the other hand, is like setting a toddler loose in a candy store; the model explores data without guidance, discovering patterns and clusters on its own. Regression models predict outcomes by examining relationships between variables, while clustering algorithms group similar data points, helping marketers target the right audience for those trendy fidget spinners.

Interactive Element: Play with an Algorithm

Here's a fun exercise: pick a mundane task you do daily, like making coffee or choosing an outfit. Break it down into a series of steps, just like an algorithm. Write each step on a sticky note and arrange them in order. Try to optimize your process. Can you skip a step or combine tasks for efficiency? Congratulations, you've just created your first algorithm! This exercise highlights the power of algorithms in simplifying complex tasks, showing how AI can streamline everyday activities by breaking them down into logical sequences.

Machine Learning vs. Deep Learning: Understanding the Difference

Picture machine learning as the diligent student in the classroom of artificial intelligence. It's that kid who reads every chapter, takes meticulous notes, and somehow always knows the answer when the teacher calls. In essence, machine learning is about developing algorithms that allow computers to learn from data. Think of it as teaching your dog a new trick: you show it the way a few times, and eventually, it gets the hang of it. This process involves two primary types: supervised and unsupervised learning.

Supervised learning is like having a teacher by your side, guiding you with examples and correcting your mistakes. It involves feeding the machine labeled data, so it can learn to predict outcomes, much like how a spam filter learns to identify unwanted emails by being shown lots of examples of spam and non-spam messages. Unsupervised learning, on the other hand, is more of a free-range approach, where the machine finds patterns on its own. It's like letting a toddler loose in a room full of toys and watching them figure out the best ones to play with. Predictive text on your smartphone is another everyday example, where the machine learns from your typing habits to suggest the next word you might want to type.

Now, let's dive into the deeper end with deep learning. If machine learning is that diligent student, deep learning is the prodigy who just knows things intuitively. It's a subset of machine learning that takes it up a notch by using artificial neural networks. These networks, inspired by the human brain, can process vast amounts of data, identifying patterns and making decisions with minimal human intervention. Imagine a self-driving car cruising down a busy street, recognizing pedestrians, traffic signals, and other vehicles, all thanks to deep learning algorithms tirelessly processing visual data.

The difference between machine learning and deep learning often boils down to complexity and scale. Machine learning can work with smaller datasets and requires more human input to refine models. It's like using a traditional map to navigate a city—effective, but needing manual guidance. Deep learning, however, thrives on big data and high computational power, relying on specialized hardware like GPUs to crunch numbers. It's more like having a GPS system that updates in real-time, constantly analyzing traffic patterns to get you to your destination efficiently.

One of the most famous deep learning feats was AlphaGo, a program developed by Google DeepMind, which defeated a professional human player in the strategy game Go, a task once thought too complex for machines. AlphaGo's success showcased deep learning's ability to handle intricate decision-making processes, far surpassing traditional game AI that relied on pre-set rules and strategies. Similarly, autonomous vehicles use deep learning to navigate and make split-second decisions on the road, ensuring safety and efficiency.

In the world of finance, machine learning models work their magic behind the scenes to detect fraudulent activities. They analyze transaction patterns, flagging anomalies that might indicate fraud. It's like having a watchful eye on your bank account, alerting you to unusual activities faster than you can say "overdraft." These models are simpler compared to deep learning systems but are highly effective for specific tasks with structured data.

In conclusion, machine learning and deep learning are like siblings in the AI family, each with its own strengths and quirks. Machine learning offers robust solutions for structured data and requires less computational power, making it accessible and versatile. Deep learning, on the other hand, shines in complex, data-heavy scenarios, offering unparalleled accuracy and adaptability. Both have found their niches, from everyday applications like spam filters to groundbreaking innovations like autonomous vehicles, proving that in the grand tapestry of AI, there's room for all kinds of learners.

Natural Language Processing: Making Sense of Human Language

Picture this: you're talking to your phone, and it not only understands your words but also responds like a slightly tech-obsessed friend. That's Natural Language Processing, or NLP for those in the know. It's the magical ability of computers to understand, interpret, and even respond to human language in a way that feels as natural as chatting with your neighbor about the weather. NLP is the bridge connecting human thoughts with machine logic, making communication with gadgets as seamless as ordering a pizza from your favorite joint—extra cheese, please. Whether it's speech recognition that lets you dictate texts while driving or language translation that turns your monologue into fluent French, NLP is hard at work, turning complex computations into everyday conveniences.

Now let's get into the nitty-gritty of how NLP works its magic. It all starts with tokenization, a fancy term for breaking down sentences into smaller pieces, or tokens. Imagine a chef chopping vegetables into bite-sized chunks before throwing them into a stew. Each word becomes a token, and this makes it easier for the machine to understand and process. Then comes parsing, which analyzes the grammatical structure of a sentence. It's like diagramming sentences back in English class, identifying subjects, verbs, and objects, so the machine knows who's doing what to whom. Sentiment analysis takes it a step further, gauging the emotional tone of the text. It's the digital equivalent of reading between the lines to figure out if a customer is praising your product or preparing to launch a complaint.

But it's not all smooth sailing. NLP faces some serious head-scratchers. Human language is filled with ambiguity, where words like "bat" can mean different things depending on whether you're talking about baseball or a creature of the night. Context plays a crucial role, much like understanding that "I'm fine" can range from genuine contentment to "I'm one snide comment away from snapping." Then there's the delightful mess of idiomatic expressions and slang. Machines have a hard time deciphering why "it's raining cats and dogs" doesn't require an ark. These challenges are what make NLP both fascinating and frustrating, as it constantly evolves to better mimic human understanding.

If you've ever asked Siri to play your favorite playlist or Alexa to turn off the lights, you've interacted with NLP in action. Virtual assistants are like the Swiss Army knives of the digital world, capable of managing calendars, setting reminders, and providing weather updates—all thanks to their NLP-powered skills. Chatbots in customer service are another prime example, fielding queries and complaints with the patience of a saint, even when users are less than polite. They use NLP to interpret requests, provide relevant responses, and even escalate issues to human agents when things get too complicated. These digital aides are transforming how businesses interact with customers, offering speedy resolutions without the dreaded "please hold" music.

In the corporate realm, NLP is the unsung hero, transforming massive amounts of unstructured data into actionable insights. It helps companies analyze customer feedback, identify trends, and tailor marketing strategies with laser precision. NLP is also making waves in healthcare, assisting doctors in sifting through patient records to provide accurate diagnoses and treatment plans. And let's not forget its contribution to accessibility, as speech-to-text technology empowers those with hearing impairments to engage with the world in new and exciting ways. NLP is not just a technological marvel; it's a powerful tool that's reshaping how we interact with the digital universe, one word at a time.

AI in Everyday Life: Real-World Examples and Applications

Imagine waking up on a cold morning. Your smart thermostat has already warmed the house to your preferred temperature while your fridge suggests a breakfast smoothie based on the dwindling supply of kale and yogurt. Welcome to a world where AI isn't just a buzzword but a part of your daily routine. Smart home devices have taken the domestic stage, turning mundane tasks into seamless experiences. It's not just about convenience; it's about transforming how we interact with our surroundings. AI-powered thermostats learn your schedule, adjusting temperatures to optimize comfort and energy use, while smart lights set the mood or even simulate your presence when you're away, keeping potential intruders at bay. Streaming services, on the other hand, have mastered the art of knowing your guilty pleasures better than you do. They use AI to sift through endless content, curating personalized recommendations that keep you glued to the screen. Whether it's the latest thriller or a heartwarming rom-com, AI ensures your next favorite show is just a click away.

In healthcare, AI is the silent partner bringing transformative changes to diagnostics and patient care. Picture a doctor examining an X-ray image, aided by AI that highlights potential anomalies with precision that rivals a seasoned radiologist. These diagnostic tools are revolutionizing medical imaging, offering faster, more accurate results that allow for timely interventions. But AI's role doesn't stop there. Predictive analytics are being used to anticipate patient outcomes, helping doctors tailor treatment plans with unprecedented foresight. It's like having a crystal ball that offers insights into a patient's future health trajectory, enabling proactive care that could save lives. The implications are profound, providing a safety net that catches health issues before they escalate.

Finance, traditionally a realm of number-crunching and risk assessment, has embraced AI with open arms. Fraud detection systems powered by AI are the unsung heroes, tirelessly scanning transactions for suspicious activities. They act as vigilant sentinels, flagging anomalies that human eyes might miss, protecting your hard-earned money from cyber threats. Meanwhile, robo-advisors are democratizing investment, offering personalized financial advice without the hefty fees of traditional advisors. They analyze market trends, personal risk profiles, and financial goals to craft investment strategies that align with your aspirations. It's like having a financial guru in your pocket, ready to guide you through the complexities of wealth management.

AI's influence extends to education, where it redefines the learning experience for students and educators alike. Imagine a classroom where each student receives personalized instruction tailored to their learning pace and style. AI platforms assess individual strengths and weaknesses, crafting a learning path that maximizes potential. It's education that listens and adapts, transforming the traditional "one size fits all" approach into a dynamic, personalized journey. Administrative tasks, too, are getting an AI makeover, with systems automating everything from attendance tracking to grading. This frees up educators to focus on what they do best: inspiring and nurturing young minds.

As we wrap up this exploration into AI's pervasive presence in our daily lives, it's clear that AI is more than a technological marvel—it's a catalyst for change across industries. From making our homes smarter to revolutionizing healthcare, finance, and education, AI offers tools that enhance efficiency, drive innovation, and empower individuals to achieve more. It's not just about keeping up with technological trends; it's about embracing a future where AI acts as a partner in our endeavors, offering insights and solutions that were once the stuff of dreams. As you continue to explore AI's potential, remember that the possibilities are as limitless as your imagination, and the journey is only just beginning.

I am going to give you prompts that you can use with AI to get you started. Keep in mind, there is no need to enter the "....." marks that are in each of the below prompts. An additional note: AI not only doesn't need you to be polite, but it actually hinders the processing of your prompts. So, please do not use things like: Please, Thank You, etc.

Smart Homes & Entertainment Prompts:

"Act as a smart home assistant. Design a daily routine for a person living in a cold climate, optimizing their heating, lighting, and food supply notifications for comfort and energy efficiency."

"Pretend you're an AI that curates TV shows and movies. Create a personalized recommendation list for someone who loves suspenseful thrillers and heartwarming romantic comedies."

"Imagine you are a home security AI. Write a plan for using lighting and smart devices to simulate an occupied house when the residents are away."

Finance Prompts:

"Pretend you are a robo-advisor. Build an investment strategy for a 35-year-old beginner investor with moderate risk tolerance and a goal to retire at 60."

"Imagine you are a budgeting AI. Help a freelance worker design a monthly savings and spending plan based on unpredictable income streams."

General 'Embrace AI in Daily Life' Exploration Prompts:

"Act as an AI personal assistant. Suggest three ways to streamline someone's morning routine using current AI technologies."

"Pretend you are a life coach powered by AI. Write a motivational message encouraging someone hesitant about using AI to enhance their career."

"Imagine you are a creative AI consultant. Suggest five unique ways an entrepreneur could incorporate AI into their daily workflow to save time and reduce stress."

How to Use These Prompts

(or, How to Talk to an Algorithm Without Sounding Like a Dalek)

Now that you've got this lovely list of AI prompts, you might be wondering what to do with them. Are they just here for decoration, like those throw pillows that no one's allowed to touch? Heavens, no. These prompts are your gateway to the good stuff — the magical, slightly uncanny experience of having a machine actually *understand* you (sort of) and then do something surprisingly useful in return.

So let's break it down. Here's how to get the most out of the prompts you've just collected.

1. Pick Your Playground

To use a prompt, you'll need a tool that responds to prompts — an AI chatbot like ChatGPT, Gemini, Claude, or anything similar. Think of it as your ever-patient, occasionally brilliant robot companion who never rolls its eyes at you.

If you're reading this on a device right now, congratulations — you're probably already halfway there.

2. Paste and Personalize

Each prompt in this book is a ready-made sentence you can copy and paste into an AI chat box. But here's the fun part: you can (and should!) tweak it.

Let's say the prompt says:

> *"Act as a robo-advisor. Build an investment strategy for a 35-year-old beginner investor with moderate risk tolerance and a goal to retire at 60."*

You could change it to:

> *"Act as a robo-advisor. I'm 42, pretty risk-averse, and dreaming of retiring on a catamaran in the Caribbean by age 58. What should I do?"*

In other words: **make it yours**. The more you tailor it to your life, your quirks, your weird dreams about retiring with parrots — the more helpful the response will be.

3. Don't Worry About Getting It Perfect

AI doesn't get offended if you're awkward or vague. You can fumble your words, misspell things, or ask follow-up questions like, "Can you explain that again but without the finance jargon?" and it will happily oblige.

It's like having a genius intern who never takes coffee breaks and is unfailingly cheerful.

4. Play, Don't Perform

Using AI should feel a bit like playing with Lego. Stack pieces together, take them apart, build something wonky, then start over. The best way to learn is by doing — and experimenting. Try odd ideas. Ask weird questions. See what happens.

You're not being graded. (Unless, of course, you *ask* it to grade you, in which case... good luck.)

5. Chain Prompts for Bigger Results

Here's a pro tip: you can feed a prompt into the AI, get an answer, and then say, "Great, now turn that into a checklist," or "Can you expand this into a weekly plan?" or even, "Write this like a pirate."

This is called **prompt chaining**, and it's as powerful as it is entertaining. It turns your AI chats into conversations — and often the best ideas show up in the second or third exchange.

6. Use AI Like a Swiss Army Knife

You'll notice these prompts are sorted by topic — smart homes, healthcare, education, and so on. But don't let that stop you from mixing them up. Ask the healthcare AI how to teach kids about nutrition. Ask the financial AI to explain budgeting using Star Wars metaphors. Go wild. The goal isn't just productivity — it's *possibility*.

7. Don't Trust It Blindly

One small (but important) note: AI is brilliant, but it can also be confidently wrong. Like a friend who always gives advice, whether or not you asked. So, double-check facts, especially with serious topics like healthcare, finances, or anything involving actual consequences.

AI is a co-pilot — not your autopilot.

Final Thought:

Learning to prompt well is a bit like learning to make pizza dough. The first one will be lumpy, the second might stick to the ceiling, but by the third or fourth, you're flinging it in the air like an old pro.

So don't worry about being fancy or technical. Be curious. Be yourself. Type like you talk. And most of all, **have fun**.

The AI is ready when you are.

Sample Conversation: Your Smart Home AI Buddy at Work

Prompt:

> *Act as a smart home assistant. Design a morning routine for someone who lives in a cold climate, likes to wake up at 6:30 a.m., and prefers coffee, natural light, and a warm bathroom floor.*

AI Response:

Certainly! Here's a cozy morning routine, smart-home style:

6:15 a.m. – Begin gently warming the bedroom to 70°F, and start preheating the bathroom floor.

6:25 a.m. – Slowly raise the smart blinds to let in natural light (or simulate sunrise with soft lighting if it's still dark outside).

6:30 a.m. – Play your preferred wake-up playlist ("Mellow Mornings") at a low volume.

6:35 a.m. – Brew coffee automatically, based on your preferred strength and roast.

6:40 a.m. – Display a brief weather and news update on your smart mirror while you brush your teeth.

6:50 a.m. – Gently cue reminders for your top three tasks of the day via smart speaker.

Would you like me to adjust this based on weekends or work-from-home days?

Why This Works:

This shows the AI responding in a structured, practical way — but it also adapts to *you*. It's specific, it's friendly, and most importantly, it doesn't feel robotic or overly technical.

You could even encourage readers to follow up with something like:

"Great! Now suggest an evening routine to help me wind down after a long day."

And voilà — they're chaining prompts like seasoned pros.

Chapter Two

Navigating AI Tools

From Basics to Mastery

IMAGINE WAKING UP ONE morning to find your to-do list magically completed, your emails sorted, and a fresh pot of coffee brewing—all thanks to your trusty AI assistant. Sounds dreamy, right? While I can't promise AI will fold your laundry or walk your dog (yet), it can certainly streamline your day-to-day life and business operations. But before diving headfirst into the vast ocean of AI tools, you need to figure out what you actually need. It's like going to an all-you-can-eat buffet; you don't want to end up with a plate full of jelly beans when you were craving a hearty steak.

Choosing the Right AI Tools for Your Needs

The first step in this AI adventure is identifying your objectives. Whether you're a business maverick hoping to automate your workflow or a curious individual looking to boost productivity, it's crucial to know what you want to achieve. Are you aiming to automate those mundane tasks that currently consume your precious coffee breaks, or are you more interested in harnessing AI for data analytics to predict your company's next big move? Perhaps you're looking to improve personal productivity or embark on a learning journey to uncover the mysteries of AI. Clearly defining your goals will help you sift through the plethora of tools available without getting lost in the digital wilderness.

Once you've pinpointed your objectives, it's time to evaluate the features of various AI tools. Look for tools with user-friendly interfaces that don't require a degree in rocket science to navigate. You want something as intuitive as scrolling through your favorite social media feed. Integration capabilities are another key factor; the tool should play well with your existing systems like an amiable office intern who gets along with everyone. Consider if the tool can seamlessly integrate with your current software stack, whether it's your CRM, email platform, or even your calendar app.

Now, let's talk about money—not the most glamorous topic, but an important one. Conducting a cost vs. benefit analysis is crucial to ensure you're getting the best bang for your buck. Some tools come with subscription models, offering regular updates and support, while others might be a one-time purchase. Free tools might catch your eye, but beware of limitations that could stifle your productivity. Premium options often provide robust features and better support, much like upgrading from a rusty bicycle to a shiny new car. Evaluate what you really need and weigh the costs against the benefits to make an informed decision.

Scalability and flexibility are your best friends when choosing AI tools that will grow with you. If you're a budding entrepreneur with dreams of expanding your business, you'll need tools that can scale as you do. Consider how the tool will adapt to your evolving needs and whether it can accommodate more users or increased data loads. Flexibility is equally important; you want a tool that will keep up with the latest AI advancements without requiring a complete overhaul every time a new update rolls out. It's like buying clothes for a growing child; you want something that can stretch and adapt as they grow. I know, that only works for a limited period of time for the kids, but with a little planning, it will go a LONG way with your AI needs.

Visual Element: AI Tool Comparison Chart

Create a simple comparison chart for three AI tools you're considering. List features such as user interface simplicity, integration capabilities, cost, scalability, and flexibility. Assign a score to each feature, then total the scores to help make your decision. This exercise will give you a clearer picture of which tool aligns best with your needs and objectives.

By carefully considering these aspects, you'll be well on your way to finding the perfect AI tools to suit your needs. Whether you're streamlining your business operations or enhancing personal productivity, the right tools can make your AI journey smoother and more rewarding.

Getting Started with ChatGPT: A Beginner's Walkthrough

Getting to grips with ChatGPT is like having a chat with a very patient friend who happens to know a little bit about everything. At its core, ChatGPT is a language model designed to understand and generate human-like text. This means it can engage in text-based conversations, making it an excellent tool for a variety of tasks. Whether you're dabbling in creative writing or looking to enhance customer service interactions, ChatGPT is your go-to digital companion. Imagine you're weaving a tale of adventure, and you hit a creative block; ChatGPT can help brainstorm new plot twists or develop quirky character backstories. In customer service, it can assist by crafting polite and informative responses, ensuring clients leave with a smile rather than a frown.

Setting up ChatGPT is a breeze, even for those who still struggle to program their TV remote. First, you create an account, which is as straightforward as signing up for your favorite streaming service. Once you're in, accessing the platform is just a click away. The initial setup involves tweaking a few basic configuration settings so that ChatGPT knows how to treat you. Do you prefer short and snappy answers, or are you the kind of person who enjoys a more detailed explanation? These settings are like setting the tone for a friendly chat over coffee, ensuring that your interactions with ChatGPT are tailored to your preferences.

Once you've set the stage, it's time to get talking. Initiating conversations with ChatGPT is as simple as typing a greeting or a question. You're not limited to formalities, either; feel free to ask it about the weather, your favorite book, or what it thinks about pineapple on pizza. When requesting specific information, clarity is key. Just like you'd ask a librarian to find a book on a particular subject, specific questions yield the best results. If you're drafting an email, for example, you might type, "Hey ChatGPT, can you help me write a professional email to my boss about taking a day off?" ChatGPT will then whip up a draft that you can tweak to perfection.

The versatility of ChatGPT shines in its practical applications. It's like having a Swiss Army knife for your digital tasks. Need to brainstorm ideas for a new project? ChatGPT can throw a dozen ideas your way, helping you sift through the noise to find the gold nugget. When you're staring at a blank screen, trying to draft an important email or document, ChatGPT steps in as your co-writer, offering suggestions that get the creative juices flowing. It's also a great sounding board for refining your thoughts, whether you're planning a presentation or jotting down notes for a meeting. And the best part? It doesn't judge your typos or the occasional overuse of exclamation marks.

Interactive Element: ChatGPT Conversation Starters

To get the most out of ChatGPT, try these conversation starters:

1. Ask it to generate a creative story prompt.

2. Request a summary of a complex topic.

3. Have it draft a friendly reminder email for a meeting.

These exercises will help you explore ChatGPT's capabilities and see how it can be tailored to your specific needs.

As you continue to explore ChatGPT, you'll discover that its potential is limited only by your imagination. Whether it's helping you write a poem, generate code snippets, or even plan your next vacation itinerary, ChatGPT is there to assist. Embrace the possibilities and see how it can fit into your daily life, making tasks a little easier and your day a bit brighter.

Exploring AI Platforms: Google AI, IBM Watson, and More

Imagine you're at a tech fair, surrounded by flashy booths each promising to revolutionize the way you do business. That's what exploring AI platforms feels like. Among the top contenders, Google AI and IBM Watson stand out like rock stars at a concert, each offering unique capabilities that cater to different needs. Google AI, with its formidable TensorFlow, serves as a powerhouse for machine learning enthusiasts. It's the go-to choice if you're looking to build complex models or dabble in neural networks without breaking a sweat. TensorFlow is as flexible as a yoga instructor, allowing you to create and train models tailored to your specific requirements. It's like having a tech-savvy friend who knows all the ins and outs of machine learning and can help you navigate the labyrinthine world of algorithms.

On the other side of the ring, we have IBM Watson, a titan in natural language processing (NLP). If words are your game, Watson is your ace in the hole. Its NLP capabilities are top-notch, making it an excellent choice for businesses that need to make sense of textual data. Whether you're wrangling customer feedback or analyzing market trends, Watson's language skills can help you turn raw data into actionable insights. It's like having a linguist at your disposal, ready to decipher the cryptic messages hidden in mountains of text. Plus, Watson's strength in data analytics means you get robust tools for understanding trends and patterns, which can be invaluable in making informed business decisions.

Choosing the right platform often feels like picking the perfect pair of sneakers—style, comfort, and functionality all matter. If your industry leans heavily on data-driven insights, IBM Watson might be your best bet with its superior analytics capabilities. It's particularly well-suited for sectors like healthcare and finance, where precise data interpretation is crucial. On the other hand, if you're in a field that demands cutting-edge machine learning applications, Google AI's TensorFlow offers the versatility and integration you need. It seamlessly fits into the Google ecosystem, which can be a boon if you're already using other Google services. The key is to align the platform's strengths with your specific needs, ensuring it complements your business objectives or personal projects.

Integrating these platforms with your existing systems can seem daunting, but it's all about finding the right balance. APIs, or application programming interfaces, are your best friends here. They act as the digital bridges connecting different software systems, allowing them to communicate effectively. Think of them as translators at a global conference, ensuring everyone understands each other despite speaking different languages. When integrating a platform like Google AI or IBM Watson, leverage APIs for data exchange, ensuring seamless interaction with your current software stack. Compatibility is crucial, so assess whether the platform can work alongside your existing tools without causing a digital traffic jam.

For those who love a good tech challenge, exploring the customization options that platforms like Google AI and IBM Watson offer can be incredibly rewarding. Personalization is key to maximizing the benefits of these platforms. Whether it's tweaking the settings to better suit your workflow or customizing the outputs to align with your brand's voice, these platforms offer a level of flexibility that can be tailored to your specific needs. It's like tailoring a suit; off-the-rack might work for some, but having something customized ensures a perfect fit. By tailoring AI platforms to your requirements, you can ensure that they serve not just as tools, but as strategic partners in your endeavors.

Navigating these AI platforms requires a bit of research and experimentation, but the potential rewards are substantial. With the right platform, you can automate processes, gain deeper insights into your data, and ultimately make more informed decisions. Whether you're a solo entrepreneur or part of a larger organization, leveraging the capabilities of AI platforms like Google AI and IBM Watson can position you for success in an increasingly tech-driven world. So, dive into the world of AI platforms with curiosity and confidence, knowing that the right tools can transform the way you work and the outcomes you achieve.

Getting Familiar with the Platforms

> "Act as a tech fair guide. Explain the differences between Google AI's TensorFlow and IBM Watson's natural language processing capabilities to someone who is new to AI, using friendly analogies."

(*Why this is good:* It gets the AI to simplify complicated stuff — and give you , the reader/explorer the opportunity to begin to experience this interaction like you might if you were chatting with a helpful friend, not taking a dry university lecture.)

> "Pretend you are a business consultant. Based on a fictional company that specializes in customer service,

recommend whether Google AI or
IBM Watson would be a better fit,
and explain your reasoning."

Hands-On with AI: Practical Exercises for Beginners

You've watched the tutorials, read the guides, and now you're wondering if it's time to put theory into practice. Well, it absolutely is! You see, hands-on practice is where the rubber meets the road in mastering AI tools. It's like cooking a new recipe—you can read about it all day, but until you actually roll up your sleeves and start chopping onions, you won't truly get it. By diving into practical exercises, you not only build confidence but solidify everything you've learned so far. It's about reinforcing your theoretical knowledge through real-world application, much like learning to ride a bike by actually hopping on one, wobbles and all.

Let's start with a step-by-step tutorial to create a simple chatbot using pre-built scripts. You don't need to be a coding wizard to get started. Think of it as assembling IKEA furniture; the instructions can be a bit tricky, but with patience, you'll have a functional piece. Begin by choosing a platform that offers pre-built chatbot templates. These are like your flat-pack furniture parts, ready to be put together with just a little guidance. Once you've selected a template, customize it with your own dialogue and responses. Imagine you're programming a digital version of yourself, ready to chat with visitors to your website or app. This exercise will have you interacting with your chatbot, tweaking its responses, and learning how to make it more engaging. It's a fantastic way to dip your toes into AI without feeling overwhelmed by complexity.

Next, let's try analyzing sentiment in social media posts. This is where AI can really shine, helping you understand what people are saying about a brand, product, or topic. Start by gathering a dataset of social media posts—this could be tweets, Facebook comments, or Instagram captions. Using an AI tool designed for sentiment analysis, run these posts through the system to categorize them as positive, negative, or neutral. It's like playing detective, uncovering insights into public opinion. You'll get to see firsthand how AI processes language and makes sense of human emotions, transforming a jumble of words into valuable data.

For those who enjoy more interactive learning, there are plenty of modules available online that offer simulations and sandbox environments. These tools let you experiment with AI in a controlled setting, much like a flight simulator for aspiring pilots. You can tweak parameters, make adjustments, and see how your changes affect the outcomes. It's a safe space to make mistakes and learn without the fear of real-world consequences. Additionally, community-driven challenges and hackathons provide a fantastic platform to test your skills. These events often come with a sense of camaraderie, as participants share knowledge and bounce ideas off one another. It's a bit like a digital science fair, where you can showcase your AI projects and pick up new tricks along the way.

Tracking your progress and milestones is just as important as getting started. Keep a journal or digital log of the exercises you've completed. Note down the challenges you faced, the solutions you found, and any aha moments you experienced. This practice is akin to keeping a workout diary when training for a marathon; it helps you see how far you've come and where you can improve. Set goals for skill improvement, whether it's mastering a new AI tool or completing a more complex project. Breaking down your learning journey into manageable milestones keeps you motivated and focused. Plus, there's immense satisfaction in ticking off those goals one by one, knowing you're steadily becoming more proficient in the world of AI.

Building a Chatbot with Pre-Built Scripts

> "Act as a chatbot-building mentor. Guide me through choosing a pre-built chatbot template for a local coffee shop, and help me customize it to greet customers, answer basic questions, and suggest today's special."

(*Super practical:* You pick a real-world business — even imaginary — and get hands-on customizing dialogue.)

> "Pretend you are a chatbot personality designer. Help me create a friendly but professional tone for a chatbot welcoming visitors to a small online bookstore."

(Think about tone, voice, and user experience — not just functionality.)

Analyzing Sentiment in Social Media Posts

> "Act as a sentiment analysis tutor. Walk me through how to gather a sample of 20 tweets about a new tech gadget, and use a free AI tool

> to categorize them as positive, negative, or neutral."

(The goal here is to turn sentiment analysis from a scary tech buzzword into a casual detective game.)

> "Pretend you are a sentiment data detective. Show me how to summarize public opinion trends from a fictional dataset of Instagram comments about a new eco-friendly clothing brand."

(Play detective here, just like you describe — and it makes the process fun, not dry.)

Troubleshooting Common AI Tool Challenges

Let's face it: technology isn't perfect. Anyone who's ever cursed at their computer screen knows that even the most sophisticated AI tools can have their off days. One moment your AI assistant is breezing through tasks like a digital superhero, and the next, it's moving slower than a snail on a lazy Sunday. Common issues like connectivity problems and lag can turn even the most patient person into a bundle of frustration. It's not just about slow responses, either. Sometimes, your AI tool might misinterpret commands, leaving you scratching your head as it generates completely irrelevant—or downright bizarre—responses. These hiccups can derail your productivity, but fear not! With a little patience and some savvy troubleshooting, you can get back on track.

Before you grab that hammer and threaten your device, take a breath and start with the basics. Think of it like giving your tech a gentle nudge to wake it up from a snooze. First, try restarting the program or device. It's amazing how often a simple reboot can solve what feels like a monumental problem. Next, check for software updates. Developers are constantly working behind the scenes to improve performance and patch bugs, so keeping your tools up to date is crucial. It's like giving your tech a shot of espresso, helping it run smoother and faster. These steps might seem basic, but they often solve the most common issues, getting your AI tools back in the game.

Of course, some problems are more stubborn than others, refusing to budge despite your best efforts. When basic troubleshooting doesn't cut it, it's time to dig a little deeper. Analyzing error logs might sound intimidating, but think of it as detective work, where you're gathering clues to solve the mystery of your tech troubles. These logs provide a glimpse into what's happening behind the scenes, offering insights into potential causes of malfunctions. If you're stumped, don't hesitate to consult online forums and support communities. They're like bustling marketplaces where tech enthusiasts share their wisdom and experiences, offering solutions and advice. Often, the answer to your problem is just a post away, hidden among threads filled with invaluable tips from those who've faced similar challenges.

But what happens when your AI tools throw a tantrum that even Sherlock Holmes couldn't solve? Sometimes, persistent system failures or integration challenges with other software require calling in the big guns. Knowing when to seek professional help is key. If your AI tool repeatedly crashes or refuses to play nice with your other software, it might be time to reach out to the experts. Consider it as calling in a tech-savvy superhero, armed with the knowledge and tools to tackle the toughest problems. They can provide tailored solutions and support, ensuring your systems work like a well-oiled machine. It's an investment that can save you time and headaches, allowing you to focus on what really matters: making the most of your AI tools.

As you navigate these challenges, remember that troubleshooting is an integral part of working with AI. It's a learning experience that not only helps solve immediate problems but also equips you with the skills to handle future tech hiccups with confidence. Embrace the process, knowing that each hurdle overcome strengthens your understanding of AI tools and their intricacies. Troubleshooting might not always be glamorous, but it's a crucial step in harnessing the full potential of AI, turning obstacles into opportunities for growth and improvement. With each challenge resolved, you're one step closer to mastering the dynamic world of AI, ready to explore new possibilities and applications.

Basic Troubleshooting and Quick Fixes

> "Act as an AI troubleshooting coach. Walk me through a simple checklist I should follow when my AI tool suddenly becomes slow or unresponsive."

(Gives you a ready-made mental flowchart — much less panic involved.)

AI Response:

Certainly! Let's walk through a few quick troubleshooting steps together:

1. **Restart the chatbot software**: Like giving it a quick nap. Many strange glitches vanish with a simple reboot.

2. **Check for recent updates**: Sometimes a new update changes how the chatbot interprets commands. If you recently updated the platform or added a new plugin, check the update notes.

3. **Review your chatbot scripts**: It's possible that a response map-

ping was accidentally edited, swapped, or corrupted. Search your script for keywords like "hours" and make sure the correct response is still linked.

4. **Run a quick test with a few basic questions**: Ask questions like "What is your name?" or "Where are you located?" to see if the odd behavior is isolated to one command or affecting everything.

5. **Check for misfired fallback settings**: Some chatbots are programmed to default to a random response when they don't understand a question. Make sure the fallback settings are working properly — not sending visitors into fruit-themed nonsense land.

If you've tried these steps and the issue persists, don't worry!I can guide you next on how to check the error logs or prepare a simple question for the support forum.

Would you like me to show you how to check your fallback settings next?

Why This Works:

- The AI breaks troubleshooting down into *bite-sized steps*, no jargon required.
- It reassures the reader — no catastrophes here, just a few sleepy settings that need tweaking.
- It invites *follow-up help*, keeping the conversation open and friendly.

Troubleshooting Tip: When things go sideways, don't think of it as failing — think of it as *finding the loose thread*. Every glitch you untangle today makes you faster and smarter tomorrow.

Chapter Three

AI for Productivity

Streamlining Daily Tasks

You wake up and—somehow—everything's already under control. Your coffee's ready, your emails have been sorted, and your calendar isn't a hot mess of overlapping chaos. And no, you didn't hire a personal assistant in your sleep. This is just life with AI pulling the strings behind the scenes.

Welcome to the new era of productivity—where your digital sidekick handles the tedious bits so you can actually breathe, think, and maybe even enjoy that first cup of coffee. Forget slogging through an inbox avalanche. With AI, you're not just surviving the day—you're steering it with surprising ease (and maybe even a little style).

Automating Routine Tasks with AI

First, let's identify which monotonous tasks are ripe for automation. Think of those repetitive chores that make you roll your eyes every time they pop up: sorting through an avalanche of emails, scheduling never-ending appointments, and the ever-dreaded data entry marathon. These are prime candidates for AI intervention. Picture AI as your personal assistant, ready to tackle these time-consuming tasks with relentless efficiency. The key is to look for tasks that follow a predictable pattern and don't require too much human creativity or judgment. After all, we're not trying to automate your next creative brainstorm session—just the mundane bits that lead up to it.

Once you've pinpointed the tasks you'd gladly hand off to a robot, it's time to choose the right AI automation tools. Platforms like Zapier and IFTTT are the unsung heroes of the digital world, ready to connect your apps and streamline your workflow like a pro. They're like digital matchmakers, helping your email app talk to your calendar without a hint of awkwardness. These tools allow you to create custom workflows that automate tasks like moving email attachments to cloud storage, adding new contacts to your CRM, or even posting reminders to your team chat. The magic lies in their ability to integrate seamlessly with your existing systems, ensuring everything runs smoother than a freshly waxed surfboard.

Now, let's dive into setting up these automated workflows. Imagine you're crafting a Rube Goldberg machine—each part triggers the next, creating a chain reaction of efficiency. Start by identifying the triggers, those little events that set off the automation. Maybe it's a new email from a specific contact or a calendar invite landing in your inbox. Once you've got your trigger, you define the actions—what you want to happen next. Perhaps the email gets sorted into a specific folder, or the appointment gets automatically added to your calendar. It's like setting up a digital domino effect, where each piece falls perfectly into place. As you set up these workflows, remember that testing and refining are part of the process. It's all about tweaking until the system runs like clockwork.

Automation doesn't stop at setting up workflows; it's an ongoing affair, much like maintaining a prized garden. You'll want to continuously monitor and optimize these processes to ensure they're running at peak efficiency. Analyze data on task completion rates to identify any bottlenecks or areas where automation might be falling short. Maybe a workflow isn't triggering as often as it should, or perhaps it's duplicating tasks like a mischievous photocopier. By keeping an eye on performance metrics, you can make necessary adjustments to keep everything humming along smoothly. Adaptation is key here; as your needs evolve, so should your workflows. It's like having a personal trainer for your digital life, pushing you to optimize and improve at every turn.

Checklist: Automation Optimization

- **Identify** repetitive tasks for automation.
- **Select** suitable AI tools like Zapier or IFTTT.
- **Set up** clear triggers and actions for workflows.
- **Test** and refine workflows for efficiency.
- **Monitor** task completion rates and adjust as needed for optimal performance.

With AI as your productivity sidekick, you can reclaim those precious hours wasted on routine tasks and redirect them toward more meaningful endeavors. Whether it's pursuing a new hobby, launching a side hustle, or simply enjoying a bit of downtime, AI is there to lend a hand, making your daily grind a little less grindy.

Finding What to Automate

> "Act as a personal productivity coach. Help me identify five daily tasks I could automate using AI to free up at least an hour each day."

> "Pretend you are an AI efficiency expert. Create a list of tasks a free-

lance writer or small business owner should automate first to maximize their time."

Choosing and Setting Up Tools

"Act as a digital matchmaker. Recommend an automation tool (like Zapier or IFTTT) for someone who wants to automatically save email attachments to Google Drive and create calendar reminders for upcoming deadlines."

"Pretend you are a Zapier setup wizard. Walk me through creating a simple workflow that saves all starred Gmail emails into a special Trello board for follow-up."

Building and Refining Workflows

"Imagine you are a workflow architect. Design a basic workflow where new customer inquiry forms trigger an automatic welcome email and a Slack notification to the sales team."

"Act as a testing coach. Help me design a checklist to test whether my automated workflows are actually working as intended before I fully rely on them."

(Trust but verify)

Monitoring and Optimizing Automations

"Pretend you are a digital personal trainer. Suggest simple metrics I should track to make sure my automations are improving my productivity, not creating more problems."

"Act as a workflow troubleshooter. If one of my automation tools stops working unexpectedly, suggest three steps I should take to quickly diagnose and fix the problem."

Dreaming Bigger with AI Productivity

> "Imagine you are a time-management magician. Design an advanced automation setup that would allow a solo entrepreneur to spend Fridays completely free from meetings, emails, and admin tasks."

(Because once you get the basics down, you'll want to build your dream life too. - From Hamster Wheel to Hammock, my first book.)

Enhancing Personal Productivity with AI Assistants

Imagine having a personal assistant who never sleeps, never takes a coffee break, and never asks for a raise. Enter the world of AI assistants—your digital sidekicks ready to tackle tasks and boost your productivity without breaking a sweat. These savvy helpers, like Google Assistant, Amazon Alexa, and Microsoft Cortana, are more than just search engines with voices. They are capable of understanding and executing tasks through simple voice commands. Whether you need a reminder to water the plants or a heads-up about an upcoming meeting, AI assistants are there to lend a hand, or rather, a voice. From setting alarms to controlling your smart home devices, they're like your digital Swiss Army knife, always ready to assist.

To get the most out of these tech marvels, you'll want to personalize your AI assistant to fit your lifestyle like a glove. Start by customizing reminders and notifications. Maybe you need a nudge to take a break and stretch those legs or a gentle reminder to call mom every Sunday. The beauty of AI assistants is that they can be tailored to your needs, turning chaos into order at the sound of your voice. Setting up daily routines is another game-changer. Picture this: you wake up, and your assistant reads the news, turns on the lights, and starts your coffee maker, all in one seamless flow. It's like having a morning routine on autopilot, giving you more time to focus on what's important—like choosing your outfit for the day.

AI assistants also shine brightly in the realm of task management. Forget about those sticky notes plastered everywhere or the never-ending lists cluttering your desk. With a few spoken words, you can create to-do lists and set reminders for everything from picking up groceries to remembering your anniversary (you definitely don't want to forget that one). They help you manage calendar events and deadlines, ensuring you never miss a meeting or double-book yourself again. It's like having a personal secretary, minus the hefty salary. Plus, with real-time syncing across devices, your tasks and reminders are always just a tap or voice command away, keeping you organized on the go.

And just when you think your AI assistant couldn't get any cooler, it can expand its capabilities through skills and integrations. Think of them as apps for your assistant, allowing it to learn new tricks and perform even more tasks. Connect it to your smart home devices, and suddenly you're living in the future where your assistant can dim the lights, lock the doors, and even adjust the thermostat with a simple command. The possibilities don't stop there. By utilizing third-party apps and services, you can teach your assistant to do everything from ordering pizza to planning your next vacation itinerary. It's like having a genie in a bottle, but instead of granting three wishes, it offers an endless array of skills to make life easier.

So, there you have it. AI assistants are more than just the voice in your smart speaker; they're the key to enhancing your productivity and simplifying your life. By configuring them to suit your needs, managing tasks with ease, and expanding their capabilities through integrations, you unlock their full potential. Whether you're coordinating your bustling schedule or simply trying to remember where you parked your car, these digital helpers are ready to assist you every step of the way.

Prompts for Enhancing Personal Productivity with AI Assistants

Personalizing Your Assistant

> "Act as a personalization coach for AI assistants. Suggest five creative, daily reminders I could set up with my AI to make my life healthier, happier, and less stressful."

> "Pretend you are an AI lifestyle designer. Help me create a personalized 'Good Morning' routine using Google Assistant or Alexa, including news, weather, coffee brewing, and a motivational quote."

Managing Tasks and Calendars

> "Act as a task manager assistant. Show me how to create a simple

voice-command system for adding grocery items, setting meeting reminders, and creating a to-do list — all synced to my phone."

"Pretend you are a productivity architect. Design a weekly planning session I could do every Sunday using my AI assistant to organize my calendar, set goals, and review upcoming tasks."

Expanding Skills and Integrations

"Imagine you are a skills trainer for AI assistants. Recommend three smart home integrations and two third-party apps that could dramatically boost my productivity at home or at work."

"Act as a smart home consultant. Help me set up a voice-activated evening routine that dims the lights, locks the doors, and plays relaxing music when I say, 'Goodnig ht.'"

Thinking Bigger: Creative Uses for AI Assistants

> "Pretend you are a digital genie. Suggest five unexpected but brilliant ways I could use an AI assistant to save time, reduce stress, or add a little more fun to my week."

AI in Time Management: Tools and Strategies

Picture this: you're staring at your overflowing calendar, wondering how on earth you're going to fit in a workout, three meetings, and a dentist appointment all in one day. Enter AI-powered time management tools, your new best friends in the quest for productivity. Tools like Clockwise and RescueTime are like personal trainers for your schedule, helping you track and optimize how you spend your days. Clockwise, for instance, intelligently moves meetings around to free up uninterrupted blocks of time, making sure you get those deep work sessions without constant interruptions. RescueTime, on the other hand, runs quietly in the background, tracking where your time goes online and gently nudging you to stay focused. It's like having a digital coach who tells you to step away from the cat videos and back to the task at hand.

Implementing AI strategies to improve time management is like giving your day a makeover, one that swaps chaos for calm. Start by analyzing your time usage patterns. These tools can show you exactly where your hours are slipping away, whether it's social media rabbit holes or endless email checking. With this insight, you can then automate your calendar scheduling. Imagine having a smart assistant who books meetings only when you're at peak productivity and blocks out "me time" for a mental recharge. This strategy not only helps you stay organized but also maximizes your efficiency, ensuring you tackle the most challenging tasks when your brain is firing on all cylinders.

Balancing work and personal life with AI might sound like a dream, but it's more attainable than ever. AI is great at setting boundaries, reminding you to step away from work to enjoy life's simple pleasures. You can set up reminders to unplug, ensuring that work emails don't invade your dinner time. AI can even help automate your work-life balance routines. Picture this: at 6 p.m., your AI assistant dims the lights, starts your favorite playlist, and gently suggests you switch off the work brain. These small automations are like friendly nudges, encouraging you to maintain a healthy balance between productivity and relaxation.

Overcoming common time management challenges with AI is like having a personal assistant who keeps you on track without the need for coffee breaks. Procrastination, for example, is a universal challenge that AI can help combat. With AI alerts nudging you to stay on task, you'll find it easier to focus and get things done. Prioritizing tasks is another area where AI shines, offering insights that help you decide what needs your attention first. Imagine having a tool that scans your to-do list and highlights the tasks that will have the biggest impact on your goals. It's like having a built-in strategist who helps you tackle your day with purpose and efficiency.

Prompts for AI in Time Management: Tools and Strategies

Analyzing and Optimizing Time Use

> "Act as a time management coach. Guide me through setting up RescueTime (or a similar tool) to track

how I spend my work hours and identify the biggest time-wasters."

"Pretend you are a focus strategist. Suggest a simple daily routine I could set up with AI tools that maximizes deep work sessions and minimizes interruptions."

Automating Calendar Scheduling

"Imagine you are a smart scheduling assistant. Help me automate my calendar so meetings are booked only during my low-energy times, and focus work is reserved for my peak hours."

"Act as a calendar optimization expert. Walk me through how Clockwise (or a similar tool) could reorganize my day to create more uninterrupted blocks of time for deep work."

Work-Life Balance and Digital Boundaries

"Pretend you are a work-life balance architect. Design a simple automation where my AI assistant reminds me to unplug at 6 p.m. and suggests a relaxing activity based on the day of the week."

"Imagine you are a digital wellness coach. Recommend three ways I could use AI to make sure work doesn't creep into my personal time — even when I work from home."

Beating Procrastination and Prioritizing with AI

"Act as an anti-procrastination buddy. Suggest a system where AI gently nudges me back on task when I spend too much time on distractions like social media or YouTube."

> "Pretend you are a priority planner. Show me how an AI tool could review my to-do list each morning and suggest which three tasks I should focus on first for maximum impact."

Sample Conversation: Using AI to Rescue a Chaotic Calendar

Prompt:

> *"Act as a smart scheduling assistant. My calendar is overloaded with meetings, random appointments, and barely any focus time. Help me reorganize my week to create larger blocks of uninterrupted work time without missing anything important."*

AI Response:

Absolutely! Let's whip that calendar into shape. Here's a step-by-step plan:

1. Identify meeting-heavy days: First, scan your current week. Are there clusters of meetings on certain days? We'll aim to bunch them even tighter, freeing up whole mornings or afternoons elsewhere.

2. Move flexible meetings: Shift any meetings that aren't urgent to later in the week or consolidate them into "meeting blocks" (example:

Tuesdays and Thursdays, 2–4 p.m.). Fewer transition gaps = more deep work time.

3. Protect your peak focus hours: Based on typical energy patterns, mornings (9–12) are usually high-focus. Let's block those out as 'Focus Time' where no meetings can be scheduled.

4. Add buffer zones: Insert 15-minute breaks between meetings where possible — to reset your brain and avoid back-to-back exhaustion.

5. Automate scheduling preferences: Use a tool like Clockwise to automatically protect focus blocks and limit meetings to designated windows.

Would you like me to suggest a draft version of what your "optimized" weekly schedule could look like? (Spoiler: it will involve less chaos, more breathing room, and maybe even time for lunch.)

Why This Works:

- The AI breaks a big scary problem (calendar chaos!) into manageable, logical steps.
- It offers **hope** and **practicality** without making the reader feel judged or overwhelmed.
- It encourages *interaction* — asking if the reader wants a draft schedule — reinforcing that *they're still in control.*

Time Management Tip: Reclaiming your schedule isn't about squeezing more in — it's about clearing space for what actually matters. (And yes, *you* matter.)

Mini Practice Missions: Put Your AI Time Manager to Work

Mission 1: Block Your First Deep Work Session

Objective: Use an AI scheduling tool (like Clockwise, Google Calendar's Focus Time, or another app) to reserve a **2-hour block** for deep, uninterrupted work this week.

Helpful Tip: Mornings are often the golden hours for focus — but you know yourself best. Pick a time when you're least likely to turn into a distracted squirrel.

Mission 2: Automate a Lunch Break Reminder

Objective: Set a daily AI reminder at your preferred lunchtime — encouraging you to *actually step away* from your desk and eat like a civilized human being.

Helpful Tip: Bonus points if you program the reminder to say something fun like, "Time to fuel up, champ!" or "Remember: your laptop doesn't taste good."

Mission 3: Declutter Your Meeting Schedule

Objective: Use an AI scheduling tool or manual review to group your meetings closer together, leaving **at least one full morning or afternoon free** of interruptions.

Helpful Tip: Meetings are like socks in the laundry — somehow they multiply if you're not careful. Group them ruthlessly and protect your quiet time like a rare endangered species.

Mission 4: Create a Work Shutdown Ritual

Objective: Set up an AI automation (through an assistant like Alexa, Google Assistant, or even just a calendar reminder) that triggers a "work shutdown" routine every day at a set time.

Helpful Tip: Your ritual could be simple: dim the lights, play chill music, close your laptop, and physically walk away. Future You will send a heartfelt thank-you note.

Mission 5: Review and Reflect

Objective: After one week of using your new AI-powered time management tools, write down:

- One thing that worked well

- One thing you'd tweak

- One small win you're proud of

Helpful Tip: Progress isn't measured in perfect days — it's measured in noticing that you're *getting better at noticing.*

Using AI to Organize Information and Data

Imagine a world where all your notes, documents, and random thoughts scattered across various devices come together into a single, well-organized repository. AI tools like Evernote, OneNote, and Notion make this dream a reality, transforming chaos into order with just a few clicks. These platforms are like digital filing cabinets, equipped with features like tagging and categorizing that make finding your notes a breeze. Evernote lets you tag your notes with keywords, so whether you're searching for last year's tax return or that genius business idea you jotted down at 3 a.m., it's all just a search away. Notion, on the other hand, offers a more structured approach to organization, allowing you to create databases that categorize information by type, priority, or project. It's like having a personal librarian at your fingertips, ready to fetch any piece of information you need.

Efficient data storage is not just about stashing away files like a digital squirrel hoarding acorns for winter. AI-driven cloud storage solutions take this concept to the next level by ensuring your data is both secure and easily accessible. Imagine having an automated backup system that saves your files to the cloud, freeing you from the dread of losing important documents to a rogue coffee spill. These systems not only store your data but also manage it intelligently, using AI algorithms to optimize storage space and speed up retrieval times. It's as if your filing cabinet could magically expand to accommodate all your documents without losing anything in the clutter. And let's not forget the beauty of automated retrieval processes, which make accessing your files feel as effortless as pulling a rabbit out of a hat.

When it comes to information retrieval, AI is like a mind reader that anticipates your needs before you even know them. Natural language search capabilities allow you to search for information using everyday language, much like asking a friend for help. For instance, instead of searching for a file named "Project Proposal Q3," you could simply type "proposal for the third quarter," and AI will do the heavy lifting to find the right document. This is complemented by predictive search features that suggest files or information based on your recent activity or frequently accessed documents. It's like having a personal assistant who knows your habits and preferences, offering suggestions that make finding what you need faster and more intuitive.

Data security and privacy are crucial aspects of managing information, and AI plays a pivotal role in safeguarding your digital assets. Imagine an AI-driven encryption system that wraps your data in layers of digital security, much like a fortress protecting a treasure trove. These encryption methods ensure that only authorized users can access sensitive information, providing peace of mind in an era where data breaches are as common as cold calls from telemarketers. Moreover, AI can automate privacy compliance checks, ensuring that your data management practices adhere to the latest regulations and standards. It's like having a watchdog that not only guards your data but also keeps track of the ever-changing landscape of privacy laws, ensuring you stay on the right side of the line.

Prompts for Using AI to Organize Information and Data

Setting Up Smart Information Hubs

> "Act as a digital organization coach. Help me set up a basic system in Notion (OneNote or Evernote) to

organize my personal notes, project ideas, and important documents into easy-to-find categories."

"Pretend you are a personal librarian. Suggest a smart tagging system I could use in Evernote to make it easy to find anything from grocery lists to business plans."

Efficient Storage and Intelligent Backup

"Imagine you are a cloud storage wizard. Walk me through setting up an automated backup system that saves important files to the cloud and keeps everything secure without me lifting a finger each day."

"Act as a file organization expert. Suggest three ways I can use AI to automatically sort new documents into folders based on content type, project, or urgency."

Smart Information Retrieval

"Pretend you are a search genie. Show me how to set up natural language search in a tool like Notion or Google Drive, so I can find documents just by describing them in plain English."

"Act as a predictive search specialist. Help me configure a system where my most-used files show up automatically based on my recent activity or common search terms."

Protecting Data and Ensuring Privacy

"Imagine you are a digital fortress designer. Recommend simple AI-driven encryption and privacy settings I should enable to keep my personal and work documents safe from prying eyes."

> "Pretend you are a privacy compliance watchdog. Help me create a checklist to make sure my document management practices stay aligned with current data protection laws."

AI-Powered Decision Making: How to Make Smarter Choices

Imagine standing at a crossroads with a pile of data as your map, and AI as your navigator, ready to guide you toward the smartest choice. Today, AI's ability to analyze vast datasets is like having a supercomputer in your corner, sifting through oceans of information in seconds. It's like speed-reading 1,000 novels at the same time, extracting only the juiciest bits to serve on a silver platter. This capability is invaluable for predictive analytics, a fancy term for forecasting future trends based on current data. It's akin to having a crystal ball, but one grounded in hard facts and statistics rather than mystical smoke. Businesses leverage this to predict market shifts, consumer behaviors, and even financial risks, making it a game-changer for strategic planning.

For those seeking to enhance decision-making prowess, a few tools stand out in the AI arena. IBM Decision Optimization is one such powerhouse, providing a framework that helps businesses make data-driven decisions with precision. It's like having a seasoned strategist on your team, calculating every possible outcome to find the optimal path forward. Meanwhile, Microsoft offers a suite of AI tools tailored for business intelligence, turning raw data into insightful reports and visualizations that even your grandma could understand. These tools are like having an in-house analyst who never sleeps, continuously churning out insights that inform and refine your strategies.

Integrating AI insights into your daily decision-making process can feel like adding rocket fuel to your strategic engine. Start by utilizing AI-generated reports for strategic planning. These reports digest complex data into bite-sized, actionable insights, providing clarity and direction in the midst of uncertainty. Imagine sitting in a meeting where, instead of guessing, you're armed with real-time data analysis that allows you to make immediate, informed decisions. It's like switching from a paper map to a GPS; the route becomes clear, and you're less likely to end up in a ditch. With AI, you can pivot quickly, responding to market changes with agility and confidence.

However, even the smartest AI can stumble, which is why evaluating AI-driven decisions for bias and accuracy is crucial. Just because a machine crunched the numbers doesn't mean it's infallible. Bias can creep in through skewed datasets or flawed algorithms, leading to decisions that favor certain outcomes over others. To counter this, it's essential to adopt techniques for validating AI recommendations. Think of it as a quality check, where human oversight ensures that the AI's output aligns with reality and ethical standards. Adjusting AI models based on feedback is another critical step, akin to tuning a musical instrument until it produces the perfect note. This iterative process helps refine the AI's accuracy, ensuring it remains a reliable decision-making partner.

In this dance of data and decisions, AI emerges as both a guide and a partner, offering clarity and foresight in a world awash with information. By leveraging AI's analytical prowess, businesses and individuals can navigate the complexities of modern life with greater ease and confidence. As we continue to explore AI's potential, remember that the power of smart decision-making lies not just in the data itself, but in the wisdom to apply it effectively, leaving room for growth, innovation, and perhaps a bit of unexpected magic.

Prompts for AI-Powered Decision Making: How to Make Smarter Choices

Understanding and Using Predictive Analytics

"Act as a predictive analytics coach. Explain in plain language how AI tools can use past data to predict future trends, and suggest one simple project where I could try this for myself."

Exploring Tools for Decision Optimization

"Imagine you are a business intelligence guide. Introduce me to IBM Decision Optimization and Microsoft AI tools, and suggest a simple beginner-friendly way to test them out."

"Act as an AI tool matchmaker. Based on someone wanting to make better personal finance decisions (like saving or investing), recommend an AI tool and explain how it would help."

Integrating AI into Daily Decision-Making

"Pretend you are my AI co-pilot. Show me how to set up a weekly habit where I review AI-generated reports or dashboards to guide my major work or life decisions."

"Act as a strategy booster. Help me create a quick checklist I can use before making decisions, where I factor in AI-generated insights alongside human judgment and gut instincts."

Validating and Adjusting AI-Driven Insights

"Imagine you are a data quality inspector. Show me three questions I should ask when reviewing AI-driven recommendations to check for possible bias or errors."

"Pretend you are a feedback loop coach. Teach me how to gather real-world results and use them to fine-tune my AI models or decision-making processes over time."

Chapter Four

AI in Business

Driving Growth and Innovation

You walk into your favorite coffee shop, and before you can say a word, your go-to latte is already waiting for you. Not because the barista is secretly psychic, but because they just *get* you. That's the kind of experience AI makes possible in modern marketing—except instead of steamed milk and espresso, it's powered by data, algorithms, and a knack for understanding human behavior.

AI dives into the nitty-gritty—every click, scroll, and purchase—so you can create customer experiences that feel less like generic ads and more like personal conversations. It's like having a marketing assistant who not only works 24/7 but also knows your audience better than they know themselves.

AI analyzes customer data in ways that would make even the most seasoned analyst envious. It delves into data, extracting nuggets of information that reveal what makes your customers tick. By segmenting and targeting, AI categorizes customers based on behaviors and preferences, allowing businesses to tailor their approaches. Whether it's knowing that Sarah prefers eco-friendly products or that John loves a good discount, AI helps create detailed customer profiles. These insights power predictive analytics, forecasting future behaviors with uncanny accuracy. Imagine knowing what your customers will crave next season, all thanks to AI's crystal ball.

Once you have this treasure trove of customer insights, you can create marketing campaigns that speak directly to your audience. AI-driven dynamic content generation allows for real-time customization, ensuring that each customer receives a tailored experience. It's like handing out personalized playlists instead of generic mixtapes, making each interaction feel special. As the campaign progresses, AI provides real-time insights that allow you to adjust on the fly, much like a DJ sensing the crowd's vibe and changing tunes accordingly. This agility ensures that your marketing efforts remain relevant and impactful, adapting to customer responses as they happen.

Enhancing customer engagement is where AI truly shines. Personalized email marketing becomes a breeze with AI tools that craft messages based on individual preferences. It's like sending love letters to your customers, each one crafted to resonate with their unique tastes. AI-powered recommendation engines in e-commerce platforms work similarly, suggesting products based on browsing history and past purchases. It's as if you have a digital personal shopper who knows exactly what you need, even before you do. These personalized interactions foster deeper connections with customers, making them feel valued and understood.

Of course, all this innovation begs the question: how do you measure the success of your AI-driven marketing initiatives? Enter AI tools designed to track and analyze customer interaction metrics. These tools offer a window into customer engagement, showing how they interact with your content and where improvements are needed. Automated reports on marketing campaign performance provide a clear picture of ROI, ensuring that your strategies are both effective and efficient. It's like having a marketing analyst who works around the clock, crunching numbers and providing insights that drive growth.

Visual Element: AI Marketing Success Checklist

- **Analyze** customer data for insights into behaviors and preferences.
- **Segment** and target audiences using AI-driven techniques.
- **Create** dynamic content tailored to individual customer profiles.
- **Engage** customers with personalized emails and recommendations.
- **Measure** success with AI tools that track and analyze interaction metrics.

With AI at your side, marketing becomes a symphony of personalized experiences, each note carefully crafted to resonate with your audience. It's about creating connections, fostering loyalty, and driving growth in ways that were once thought impossible.

Revolutionizing Customer Service with AI Chatbots

Imagine having a customer service team that never sleeps, never takes a coffee break, and never calls in sick. Welcome to the world of AI chatbots, where customer service is available 24/7, ready to handle inquiries and complaints at a moment's notice. AI chatbots are like the friendly, always-alert concierge at a luxury hotel, eager to assist with anything from booking reservations to providing recommendations. These digital assistants transform customer service by offering instant support, ensuring that no question goes unanswered and no customer leaves frustrated. They tackle common inquiries with ease—whether it's checking the status of an order or resetting a forgotten password—saving human agents for more complex issues that require a personal touch.

Integrating AI chatbots into your business isn't as daunting as it seems. First, you'll need to select the right platform, one that aligns with your business needs and customer expectations. It's like choosing the perfect partner for a dance—compatibility is key. Look for a chatbot platform that offers customization, allowing you to tailor responses and interactions to reflect your brand's voice. Once you've chosen your platform, the next step involves training the chatbot with business-specific data. Think of it as teaching a new employee the ropes. Feed the chatbot with information about your products, services, and common customer queries. This data acts as the foundation upon which your chatbot builds its knowledge, enabling it to provide accurate and relevant responses.

AI chatbots are not just about efficiency; they're also about enhancing customer satisfaction. By personalizing interactions based on user history, chatbots create experiences that feel tailor-made. It's like walking into your favorite café and having the barista remember your usual order. Chatbots analyze past interactions, offering personalized recommendations and solutions that resonate with individual customers. Additionally, chatbots break down language barriers with multilingual support, allowing businesses to engage with a global audience. This capability ensures that customers from different regions feel valued and understood, regardless of the language they speak.

To ensure your chatbot is hitting the mark, it's crucial to evaluate its performance regularly. Metrics for response accuracy and speed provide insights into how well the chatbot is handling customer interactions. It's like having a report card that highlights areas of excellence and those in need of improvement. Analyzing these metrics helps identify patterns and pinpoint issues that may require tweaks. Customer feedback and satisfaction surveys are equally important, offering direct insights into how customers perceive their interactions with the chatbot. This feedback acts as a guiding light, helping refine and optimize the chatbot's performance for better service delivery.

Interactive Element: Chatbot Evaluation Checklist

- **Select** a chatbot platform that aligns with your business needs.
- **Train** the chatbot with business-specific data for accurate responses.
- **Monitor** response accuracy and speed for performance insights.
- **Gather** customer feedback to refine and optimize the chatbot experience.

AI chatbots revolutionize customer service by offering instant, personalized support that enhances satisfaction and streamlines operations. They're the tireless, efficient team members that businesses need to stay competitive in a fast-paced world.

Prompts for Revolutionizing Customer Service with AI Chatbots

Choosing and Setting Up a Chatbot

> "Act as a chatbot consultant. Help me choose an AI chatbot platform that would work best for a small business focused on online retail, offering easy customization and good customer support."

"Pretend you are a chatbot onboarding specialist. Walk me through the first three things I should teach my new AI chatbot to ensure it answers common customer questions correctly."

Training and Customizing the Chatbot

"Imagine you are a chatbot trainer. Suggest a beginner-friendly way to feed my chatbot business-specific data, like product FAQs and common troubleshooting questions."

"Act as a brand voice coach. Help me customize my chatbot's tone and style so it matches a casual, friendly brand personality, rather than sounding too formal or robotic."

Personalizing the Customer Experience

"Pretend you are a personalization strategist. Show me how an AI chatbot could use past customer interactions to offer smarter, more personalized recommendations."

"Imagine you are a global support consultant. Suggest simple ways to add multilingual capabilities to my chatbot so I can serve international customers more effectively."

Evaluating and Improving Chatbot Performance

"Act as a performance analyst. Help me create a simple checklist to monitor my chatbot's response accuracy, speed, and customer satisfaction over time."

"Pretend you are a customer feedback expert. Design a short feedback form I could offer at the end of a chatbot conversation to gather insights on what's working and what needs improvement."

AI for Financial Forecasting and Analysis

Picture this: You're in a bustling stock exchange, traders shouting numbers like it's a high-stakes auction. But instead of joining the chaos, you're sipping coffee in a quiet office, watching as AI does the heavy lifting. In the world of finance, AI is like the ultimate data cruncher, sifting through mountains of information faster than you can say "bull market." Its ability to process vast datasets with lightning speed is a game-changer for financial forecasting and market analysis. By using predictive modeling, AI identifies trends and patterns that even the keenest human analysts might miss. Imagine having a crystal ball that doesn't just show the future but also explains why it's happening. That's AI in finance—turning raw data into actionable insights, making it easier to forecast economic shifts and market movements with confidence.

Now, let's talk about risk management, because let's face it, nobody likes surprises when it comes to their money. AI steps in as the vigilant guardian, identifying potential financial risks with its sophisticated algorithms. These algorithms are like hawk-eyed sentinels, scanning for irregularities and potential threats in real-time. Whether it's detecting unusual transaction patterns or flagging suspicious activities, AI ensures that nothing slips through the cracks. This real-time monitoring is crucial for mitigating risks, allowing financial institutions to react swiftly and decisively. It's like having a financial bodyguard who never sleeps, constantly on the lookout for trouble while you focus on strategizing your next move.

When it comes to optimizing investment strategies, AI is the savvy advisor you never knew you needed. AI-driven portfolio management tools are transforming how investors approach the market. These tools analyze a plethora of data—from historical trends to current market conditions—crafting strategies that are both precise and adaptable. It's like having a seasoned financial guru whispering the best moves into your ear. Machine learning models take it a step further, predicting stock market fluctuations with remarkable accuracy. These models learn from past data, adjusting their predictions as new information becomes available. It's akin to playing chess with a computer that not only knows the rules but anticipates your every move.

In the realm of financial reporting, AI is the unsung hero, automating processes that once required painstaking manual effort. Imagine generating comprehensive financial reports with a few clicks, reports that are accurate, detailed, and ready for presentation. AI handles the heavy lifting, compiling data from various sources and presenting it in a coherent format. This automation not only speeds up the reporting process but also significantly reduces the chance of human error. No more squinting at spreadsheets or worrying about misplaced decimals—AI ensures that every figure is in its rightful place. Automated data entry further streamlines the process, allowing financial professionals to focus on analysis and strategy rather than mundane data entry tasks. It's like having a diligent assistant who never misses a beat, ensuring that your financial reports are always polished and precise.

With AI at the helm, financial forecasting and analysis become more than just numbers on a page. They transform into dynamic, insightful narratives that guide decision-making and strategy. Whether you're managing a portfolio, mitigating risks, or crafting financial reports, AI offers tools and insights that empower you to navigate the complex world of finance with confidence and clarity.

Prompts for AI for Financial Forecasting and Analysis

Predictive Modeling and Forecasting

"Act as a financial forecasting coach. Explain how AI uses predictive modeling to forecast stock trends, and suggest a beginner-friendly project I could try to see predictive analytics in action."

"Pretend you are an AI financial strategist. Show me how I could use free or affordable AI tools to predict market movements or trends in a simple industry like retail or technology."

Risk Management and Threat Detection

"Imagine you are a risk management watchdog. Teach me how AI tools can scan for suspicious transactions or investment risks, and suggest a basic setup for personal finance or small business accounts."

"Act as a digital financial bodyguard. Help me set up basic AI alerts for unusual spending patterns or stock price drops in my investment apps."

Optimizing Investment Strategies

"Pretend you are a portfolio optimization advisor. Walk me through how AI can help build a diversified investment portfolio based on risk tolerance and financial goals."

"Imagine you are a chess master of investments. Suggest how machine learning models can improve a basic stock-picking strategy over time."

Automating Financial Reporting and Data Entry

"Act as a financial reporting assistant. Show me how AI can automate monthly financial report generation for a freelance business or a small company."

"Pretend you are a data accuracy wizard. Help me set up a basic system where AI double-checks my spreadsheets for errors before finalizing reports."

Supply Chain Optimization: AI Applications

Imagine standing in a massive warehouse, surrounded by towers of boxes all waiting for their ticket to ride on the logistics express. It's a chaotic scene, but AI steps in like a calm, efficient conductor, orchestrating a symphony of supply chain perfection. In the world of supply chain management, AI is the unsung hero, optimizing everything from inventory management to demand forecasting. It's like having a crystal ball that tells you exactly how much stock you need and when, preventing the dreaded surplus or shortage. AI uses historical data and predictive analytics to anticipate demand, ensuring shelves are stocked just right. No more guessing games or last-minute scrambles to meet customer needs; instead, you have a streamlined process that feels almost magical in its precision.

When it comes to logistics and transportation, AI is like the ultimate traffic controller, ensuring everything runs smoothly and on time. It analyzes routes, predicts traffic patterns, and optimizes delivery schedules so that shipments arrive at their destination without a hitch. Imagine a world where delivery trucks glide through traffic as if guided by an invisible hand, arriving precisely when expected. AI not only reduces delays but also minimizes transportation costs, making logistics a well-oiled machine. By streamlining these processes, businesses can save both time and money, delivering products more efficiently and enhancing customer satisfaction.

Improving supplier relationships is another area where AI shines, acting like a savvy negotiator at the table. It enhances interactions by analyzing supplier performance and offering insights that help businesses make informed decisions. AI examines delivery times, quality metrics, and cost-effectiveness, providing a comprehensive view of supplier performance. This information is invaluable for maintaining strong partnerships and ensuring that suppliers meet their commitments. Additionally, AI automates negotiation and contract management, allowing businesses to navigate complex agreements with ease. It's like having a digital diplomat who ensures that all parties are on the same page, fostering collaboration and mutual growth.

Operational efficiency gets a significant boost with AI, which reduces lead times and enhances productivity. AI scheduling tools are like expert timekeepers, ensuring that every process is perfectly timed to minimize waiting and maximize output. They analyze production schedules and adjust them dynamically, keeping everything running like a Swiss watch. Predictive maintenance for supply chain equipment is another ace up AI's sleeve, using data to anticipate when machines might need a tune-up. This proactive approach prevents unexpected breakdowns and costly downtime, ensuring that the supply chain hums along without a hitch.

Achieving real-time supply chain visibility is the cherry on top, allowing businesses to monitor shipments and deliveries with pinpoint accuracy.

AI-powered tracking systems provide a bird's-eye view of the entire supply chain, offering insights that are as clear as a sunny day. With anomaly detection, AI alerts businesses to any irregularities or disruptions, enabling swift corrective action. It's like having a vigilant guardian overseeing operations, ensuring that everything stays on track. This level of visibility empowers businesses to respond to challenges quickly, maintaining the integrity and efficiency of their supply chain operations.

Prompts for Supply Chain Optimization: AI Applications

Inventory Management and Demand Forecasting

"Act as a supply chain forecaster. Help me set up a basic AI-driven system to predict product demand based on historical sales and seasonal trends."

"Pretend you are an inventory optimization coach. Suggest a simple strategy using AI to maintain the perfect balance of inventory — not too much, not too little — for a small online business."

Logistics and Transportation Efficiency

"Imagine you are a logistics traffic controller. Show me how to use AI tools to optimize delivery routes for faster shipping times and lower fuel costs."

"Act as a delivery optimization expert. Help me design a basic AI system that predicts delays based on weather, traffic, or supply disruptions and adjusts delivery schedules accordingly."

Supplier Relationship Management

"Pretend you are a supplier relationship coach. Walk me through how AI can analyze supplier performance and suggest improvements to strengthen partnerships."

"Imagine you are a contract negotiation assistant. Show me how AI can help automate parts of supplier contract management to make negotiations faster and more accurate."

Operational Efficiency and Predictive Maintenance

"Act as a production efficiency strategist. Help me set up a basic AI scheduling system that minimizes downtime and maximizes output for a small manufacturing team."

"Pretend you are a predictive maintenance wizard. Explain how I could use AI sensors and data analytics to prevent machine breakdowns in a warehouse or factory setting."

Real-Time Supply Chain Visibility

"Imagine you are a real-time supply chain tracker. Help me create a simple system where I can see the status of all my shipments in one dashboard, with AI alerts for any problems."

"Act as a disruption detective. Teach me how anomaly detection in AI can help me spot and fix supply chain problems before they snowball into bigger issues."

AI in Product Development: From Concept to Market

Picture walking into a room filled with endless possibilities, where every idea has the potential to become the next big thing. This is the world of product development, and AI is the secret ingredient turning concepts into reality faster than you can say "innovation." When it comes to product design, AI tools act like seasoned designers, helping you refine ideas and bring them to life. Imagine AI-assisted prototype development, where virtual models are created and tested with a precision that would make even the most meticulous engineer proud. These tools simulate real-world conditions, allowing you to tweak and perfect your design before a single physical prototype is made. It's like having a crystal ball that shows how your product will perform in real life, saving time and resources.

But it doesn't stop there. AI also plays a crucial role in user feedback analysis, providing insights that are more insightful than a focus group of seasoned critics. AI models sift through user data, identifying patterns and preferences that might not be immediately obvious. This analysis helps refine designs to better meet user needs, ensuring that the final product is not just functional but also user-friendly. It's like having a personal coach who knows exactly what tweaks will make your product a hit, guiding you toward a design that resonates with users on a deeper level.

Accelerating product development cycles is another area where AI shows its prowess. By automating repetitive design tasks, AI frees up time for creativity and innovation. Imagine an AI tool that handles the mundane tasks of resizing images or aligning text, allowing designers to focus on the big picture. These tools work tirelessly, ensuring that no detail is overlooked and that the project moves forward without unnecessary delays. AI-driven project management tools also keep everything on track, much like a project manager who never sleeps, coordinating tasks, setting deadlines, and ensuring that everyone is on the same page. This streamlined approach reduces time-to-market, allowing you to launch products faster and stay ahead of the competition.

When it comes to enhancing product quality, AI is the vigilant inspector that never misses a flaw. AI-powered inspection systems perform quality control with laser-like precision, identifying defects and inconsistencies that might escape the human eye. Imagine a system that scans each product component with the thoroughness of a detective on a high-profile case. Real-time defect detection and correction mean that issues are addressed promptly, ensuring that every product meets the highest standards before it reaches the consumer. It's like having an army of perfectionists ensuring that your product is nothing short of exemplary.

A successful product launch is the grand finale, and AI is the maestro conducting the orchestra. AI tools for market analysis and consumer trend prediction provide valuable insights that guide the launch strategy. They analyze market conditions, consumer behaviors, and competitor activities, providing a roadmap for a successful debut. Post-launch, AI continues to monitor product performance, offering insights that help fine-tune marketing strategies and address any issues that arise. It's like having a backstage pass to the product's life, ensuring that it continues to thrive long after the initial fanfare.

In this fast-paced world of product development, AI acts as both a guide and a partner, transforming ideas into tangible, market-ready products. It's about bringing innovation to life, ensuring quality, and navigating the complexities of the market with confidence and foresight. As you continue to explore AI's potential, remember that the possibilities are as limitless as your imagination, and the journey is only just beginning.

Prompts for AI in Product Development: From Concept to Market

AI in Product Design and Prototyping

> "Act as a product design coach. Help me use AI to create a basic prototype for a new kitchen gadget, testing it virtually for usability and durability before making a physical model."

> "Pretend you are an AI design consultant. Suggest three beginner-friendly AI tools I could use to turn a rough idea into a 3D virtual prototype for testing."

User Feedback Analysis

"Imagine you are a user feedback detective. Show me how I could use AI to analyze customer reviews and user data to improve the design of an existing product."

"Act as a user experience coach. Help me create a simple system where AI highlights patterns in user feedback so I can spot recurring design flaws or feature requests."

Accelerating Product Development Cycles

"Pretend you are a creative time-saver. Suggest ways I can automate repetitive design tasks (like resizing images or checking formatting) so I can focus more on innovation."

"Imagine you are a project management AI. Help me set up a simple AI-powered project tracker that keeps my product development timeline on schedule without micromanagement."

Enhancing Product Quality

"Act as a quality control inspector. Show me how AI can be used to automatically detect defects or inconsistencies in product parts before they ship to customers."

"Pretend you are a product testing strategist. Suggest simple ways I could use AI to test product durability or reliability virtually before mass production."

Launching and Monitoring Products

"Imagine you are a market analyst. Walk me through how I could use AI to predict the best time and strategy for launching a new eco-friendly product."

"Act as a post-launch performance coach. Help me set up a system where AI monitors product reviews, sales trends, and customer support issues after launch, giving me regular reports."

Chapter Five

Prompt Engineering

Mastering AI Interaction

Picture this: you're at your favorite coffee shop (yes, I like coffee, although I typically just drink it black), and instead of just ordering your usual cappuccino, you suddenly think, "What if I could ask the barista to whip up something unique, like a caramel macchiato with a hint of cinnamon and a sprinkle of mystery?" That's kind of how prompt engineering works with AI. It's about crafting the perfect request to get exactly what you want, only instead of coffee, it's information or tasks from an AI model. Prompt engineering is the art and science of writing prompts that guide AI to deliver the output you're dreaming of. It's like being the director of a movie, where the script you write determines the scene that unfolds.

At its core, a prompt is an input—a series of instructions or questions you provide to an AI model to steer its responses. Think of it as giving directions to a friend who's lost on their way to your house. The more precise your directions, the sooner your friend arrives without getting distracted by every interesting detour. Simple prompts might look like, "Tell me a joke," while complex ones could be, "Describe the impact of quantum computing on modern cryptography." The complexity of the prompt determines the depth and specificity of the AI's response. You wouldn't ask a five-year-old to explain the theory of relativity, and similarly, prompts need to be tailored to what the AI is capable of understanding and producing.

Understanding how AI interprets these prompts is crucial to mastering AI interactions. AI doesn't possess human intuition or emotion; it relies on patterns and context to deliver responses. This means that the more context and specificity you provide, the clearer and more relevant the AI's output will be. Imagine asking a chef to make dinner without specifying any preferences—you might end up with a dish you can't pronounce or even eat. Context acts as the seasoning, enhancing the flavor of the AI's response. By setting the scene with detailed prompts, you guide the AI to generate answers that are not only accurate but also aligned with your expectations.

However, crafting effective prompts comes with its own set of pitfalls. One common mistake is using overly vague or ambiguous prompts. It's like telling someone to "be creative" without any direction. The AI, much like a person, can flounder without a clear path, leading to responses that are generic or off-target. Another issue is failing to account for the AI's limitations. While AI is a powerful tool, it's not infallible. It can't read minds or infer intentions without being explicitly told what you want. Assuming it knows more than it does often results in disappointing interactions.

To avoid these pitfalls, it's important to establish clear objectives for your prompts. Start with a solid understanding of what you want to achieve. If you're using AI to generate marketing content, for instance, your prompt should align with the desired tone, audience, and message. This goal-oriented approach keeps the interaction focused, much like a well-written to-do list. By setting clear intentions, you help the AI understand what success looks like, making it easier for it to deliver results that meet your needs. A prompt like, "Create a friendly, engaging product description for a new eco-friendly water bottle aimed at young adults," clearly outlines the task and desired outcome, guiding the AI model to craft a response that hits the mark.

Interactive Exercise: Prompt Crafting Practice

Take a moment to craft a prompt for an AI tool of your choice. Focus on a task you frequently encounter, such as drafting an email or generating a creative story idea. Write your initial prompt, then refine it to include more context and specificity. Compare the AI's responses to see how clarity and detail can enhance the output. This exercise will help you hone your prompt engineering skills, turning you into a maestro of AI interactions.

Crafting Effective Prompts for AI Tools

Imagine you're trying to get your AI to whip up a snazzy marketing slogan, and it churns out something like "Buy Stuff Now!" Clearly, it's time to polish those prompts. The magic of a well-crafted prompt lies in its three musketeers: clarity, specificity, and relevance. You wouldn't ask a chef to cook "something tasty" and expect a gourmet meal, right? Just as you'd specify "a spicy chicken curry with a hint of lime," your prompts should leave no room for guesswork. This means using precise language that conveys exactly what you want, avoiding vague terms that could send the AI down a rabbit hole of confusion. Picture yourself as a lighthouse guiding AI ships through the fog of ambiguity; your light (or prompt) should be bright enough to show the way without any detours.

Structuring prompts isn't just about what you say, but how you say it. A little structure goes a long way in making the AI's life easier, and numbered lists or bullet points can be your best friends here. It's like handing the AI a to-do list rather than a stream-of-consciousness monologue. For example, if you're asking an AI to draft an email, break it down: 1) Greet the recipient, 2) Introduce the main point, 3) Provide details, 4) Conclude with a call to action. Adding context within your prompt is equally crucial. It's akin to setting the stage before an actor steps into the spotlight. If the AI knows your email is about a new product launch, it's more likely to hit the right tone and content, much like an actor delivering a line with the right emotion.

Crafting prompts is a bit like sculpting; you start with a rough block and refine it until it takes shape. Iteration is your chisel, and each version of your prompt brings you closer to the masterpiece. Begin with a draft and test it out. Does the AI deliver what you expect, or does it wander off into the weeds? If the latter, tweak your prompt. Maybe add more detail or clarify your request. It's an evolving process, where each tweak is a step toward perfection. Think of it as training a puppy; patience and small adjustments lead to the best results. Over time, you'll find that your prompts become sharper, more effective, and tailored to coax out the best from your AI companion.

Let's look at some stellar examples of effective prompts that have proven their mettle. In customer service chatbots, prompts like "How may I assist you with your inquiry today?" often set the stage for successful interactions. They're open-ended enough to invite the customer's input but framed to guide the conversation naturally. For eliciting detailed AI-generated responses, try prompts like "Explain the benefits of using energy-efficient appliances in a home setting," which give the AI a clear topic and direction. These examples illustrate the power of specificity and context, showing how a well-thought-out prompt can transform a simple interaction into a meaningful exchange. The right prompt not only guides the AI but also ensures that the response is relevant, informative, and aligned with your goals.

Example of a Poorly Crafted Prompt

Imagine handing this prompt to your AI:

> "Write something cool about stuff
> people might like."

Sounds casual and easygoing, right? But here's the problem: the AI is standing there blinking at you like a confused puppy, wondering, *"Cool how? About what stuff? For whom? Are we talking about sneakers? Tax law? Alpacas?"*

This prompt is **too vague**, **too broad**, and **lacks context**. It's like asking a stranger for "something good" at a restaurant without mentioning you're allergic to shellfish — you're probably not going to love what you get.

When you don't give the AI clear guidance, it has to guess — and the results are often hilariously (and frustratingly) off the mark. You might get an article about popular hobbies, or a list of best-selling gadgets, or a random motivational quote ... all technically about "stuff people might like," but none of it actually useful for your needs.

Why This Prompt Fails:

- **No Specific Topic**: "Stuff" could mean *literally anything*.
- **No Defined Audience**: Who are we writing for? Teenagers? CEOs? Retired skydivers?
- **No Clear Goal or Format**: Should it be a blog post? A tweet? A jingle?
- **No Context or Constraints**: Tone, length, focus — it's all left wide open.

Quick Fix: A Better Version

Instead of the vague original, try something like:

> "Write a 50-word marketing slogan for a new eco-friendly reusable water bottle targeted at college students. Keep the tone upbeat and modern."

Now the AI has everything it needs: **Topic, audience, purpose, tone, and format.** You're not just tossing the AI a fishing rod and hoping it figures it out — you're handing it a map, bait, and pointing straight to the best part of the lake.

Interactive Exercise: Prompt Refinement

Take a prompt you frequently use and write it down. Now, imagine this prompt is an employee under review. How can it improve? Add specificity, structure, and context, then test it with your AI tool. Compare the before and after responses to see how refinement can enhance the AI's output. This exercise will help fine-tune your skills, turning you into a prompt wizard capable of getting the most out of your AI tools.

Advanced Techniques in Prompt Engineering

Imagine for a moment you're a detective piecing together clues at a crime scene, only your crime scene is a digital page, and your clues are words. Contextual cues are your magnifying glass, sharpening the focus of AI responses. When you provide background information within prompts, you're giving AI a roadmap, ensuring it doesn't take a wrong turn. Think of writing an AI-generated article on the Industrial Revolution. By including historical context, such as the invention of the steam engine or the impact on urbanization, you're equipping the AI with the right tools to craft an insightful narrative. It's like leaving breadcrumbs that lead to a treasure trove of accurate and relevant content, minimizing the risk of the AI going rogue with unrelated facts.

Now, let's talk about feedback loops. Picture them as your personal editors, helping refine your prompts to perfection. When you collect AI responses and analyze them, you gain insight into how well your prompts are performing. It's a bit like taste-testing your homemade soup; a little too salty, you adjust, a bit bland, you add seasoning. This process of continuous refinement ensures your prompts keep getting better and better. If a prompt produces unexpected or lackluster results, tweak it and try again. It's an iterative dance that polishes your prompts until they shine, harnessing the AI's potential more effectively with each go. Over time, this practice becomes second nature, a trusty tool in your AI toolkit.

Imagine AI as a decision-making maestro, wielding the power of conditional logic. This technique is your sheet music, guiding AI through complex scenarios with finesse. By using if-then prompts, you set conditions that dictate AI responses based on specific criteria. Suppose you're using AI for customer support: "If the customer mentions a billing issue, then provide the billing department contact." It's like programming a virtual assistant to handle varied scenarios without missing a beat. Conditional logic ensures the AI doesn't go off-script, maintaining consistency and professionalism in every interaction. It's a clever way to keep the AI on track, ensuring it knows exactly what to do when faced with multiple possibilities.

Let's consider prompt chaining, a technique that transforms AI into a multitasking wizard. By breaking down complex tasks into a sequence of prompts, you guide AI through each step like a well-rehearsed choreography. Imagine creating a step-by-step guide for assembling a piece of furniture. Instead of overwhelming AI with a single, massive prompt, you guide it through each stage: identifying parts, laying out tools, and following assembly instructions. Each prompt builds on the last, creating a logical flow that enhances accuracy and clarity. Prompt chaining is like assembling a puzzle piece by piece, revealing the bigger picture as each component falls into place. This approach is especially useful for intricate tasks that require precision and detail, ensuring the AI's output is coherent and comprehensive.

Example: Here's a simple example using prompt chaining to **write a short story**:

Step 1: Generate a Story Idea

Prompt: *Give me a unique story idea involving a time traveler and a forgotten city.*

Output: *A time traveler from the year 2475 discovers a city buried under the Sahara Desert that predates all known civilizations. As he explores, he realizes the city is his own creation, sent back through time in a paradox.*

Step 2: Expand the Plot

Prompt: *Expand this story idea into a 3-paragraph summary with a beginning, middle, and end.*

Input: *A time traveler from the year 2475 discovers a city buried under the Sahara Desert that predates all known civilizations. As he explores, he realizes the city is his own creation, sent back through time in a paradox.*

Output:[3-paragraph plot summary describing how he finds the city, investigates its secrets, and eventually realizes he's part of the loop that created it.]

Step 3: Write the Opening Scene

Prompt: *Write the opening scene of this story, based on the summary you just gave me. Make it mysterious and atmospheric.*

Input: [3-paragraph summary from Step 2]

Output: [A vivid opening scene where the time traveler emerges from his pod and stumbles upon the glinting spires of a forgotten city beneath the sand.]

You can chain this even further by asking the AI to:

- Develop characters from the scene
- Add dialogue
- Write a twist ending
- Turn the story into a screenplay

Tailoring Prompts for Specific AI Tasks

Imagine you're a chef with the task of cooking up a storm using an assortment of mysterious ingredients. The success of your culinary creation depends on your ability to tailor the recipe to suit the dish you're trying to create. Similarly, when crafting prompts for AI, you need to customize them based on the specific AI application at hand. In the realm of creative writing, prompts might take on a whimsical tone, encouraging the AI to weave a narrative with a dash of flair and imagination. A prompt like "Describe a world where cats rule and dogs drool" can unlock a treasure trove of creativity, sparking responses that surprise and delight. On the flip side, if you're dealing with technical documentation, your prompts need to be precise and focused. A prompt such as "Generate a step-by-step guide for calibrating a 3D printer" requires clarity and structure to ensure the AI delivers information that's both accurate and accessible. This tailoring process is like adjusting your wardrobe for a change in seasons—adapt the style to fit the task, and you'll achieve the results you're aiming for.

Adapting prompts for different AI models is a bit like figuring out which tools to use for a home improvement project. You wouldn't use a hammer to tighten a screw, and the same principle applies here. Different AI models come with their own sets of abilities and limitations, much like a Swiss Army knife with individual tools for specific needs. Some models excel at language generation, while others might be better suited for data analysis or image recognition. Understanding these differences allows you to tailor your prompts accordingly, ensuring that you're asking the right questions of the right AI. For instance, a language model adept at storytelling might stumble over a complex mathematical query, while an AI trained in technical analysis will handle it well. Adapting your approach based on the AI model's strengths ensures you get the best possible outcome for the task at hand.

Scenario-based prompt crafting is your secret weapon for tackling challenges that pop up like unwanted weeds in a garden. By envisioning specific scenarios, you can craft prompts that guide the AI through a labyrinth of possibilities with precision and finesse. Consider crafting prompts for AI-driven educational tools. You might design a scenario where the AI acts as a virtual tutor, ready to explain complex calculus concepts to a student struggling to grasp the basics. A prompt like "Explain the concept of derivatives using real-world examples and analogies" sets the stage for a tailored interaction that meets the student's needs. In other scenarios, such as customer support, prompts can be structured to guide the AI through potential issues a customer might face, ensuring they receive timely and relevant assistance. It's a bit like choreographing a dance; each step is carefully planned to ensure the overall performance is both seamless and effective.

Evaluating the effectiveness of prompts across different tasks is akin to running a taste test for your latest culinary creation. You need to assess how well your prompts are performing and make adjustments where necessary. Metrics for prompt performance evaluation can include variables such as response accuracy, user satisfaction, and task completion rates. By analyzing these metrics, you gain insights into what's working and what needs tweaking, much like adjusting the seasoning in a recipe. Feedback collection from AI tool users is equally valuable—a direct line to the end-users who can reveal how the AI's responses are perceived. Their input acts as a compass, guiding you toward improvements that enhance the overall interaction. This iterative process is like refining a recipe over time, ensuring each iteration is better than the last. As you refine your prompts based on these evaluations, you unlock the full potential of AI, transforming it into a reliable partner in your endeavors.

Case Studies: Successful Prompt Engineering Examples

Let's step into the shoes of a healthcare professional juggling patient records and treatment plans. In this industry, AI with well-crafted prompts has become a game-changer. Take the case of a hospital that implemented AI to manage patient data. By designing prompts that directly addressed patient symptoms and medical history, the AI could provide doctors with treatment options faster than you can say "hypochondriac." This not only reduced wait times but also improved diagnostic accuracy, giving doctors more time to focus on patient care rather than drowning in paperwork. Similarly, in the finance sector, banks commonly use AI to enhance fraud detection. By crafting prompts that analyze transaction patterns and flag anomalies, they are able to identify fraudulent activities with stunning accuracy. This not only saves money but also builds customer trust, as the bank could assure clients their assets are safe and sound.

Education offers another fascinating arena where prompt engineering shines. Imagine a classroom where AI helps tailor lessons to each student's learning pace. A school district implemented AI to assist teachers in creating personalized learning experiences. Prompts were designed to analyze student performance, identifying areas where a student struggled. This allowed teachers to offer targeted support, boosting student confidence and overall performance. It was like having a personal tutor for each student, without the hefty price tag. These examples showcase how effective prompts lead to better outcomes, not just in efficiency but also in enriching the user experience.

On the flip side, not every attempt at prompt engineering hits the mark. Consider a tech startup that tried using AI to generate customer service responses. They went all in, thinking AI would handle everything seamlessly. However, they soon realized their prompts were too broad, leaving customers more confused than a cat chasing its own tail. The AI spat out responses that were either too generic or completely off-base. The lesson here? Specificity is your best friend. They learned to refine their prompts, narrowing them to specific scenarios, which vastly improved the AI's performance. Another common pitfall is assuming the AI can infer context without being explicitly told. This often leads to responses that miss the mark, leaving users scratching their heads. By understanding these failures and addressing them head-on, businesses can refine their approach, turning missteps into stepping stones for improvement.

Prompt engineering doesn't just improve traditional business metrics; it can spark creativity in unexpected ways. Take the world of art and entertainment, where AI is being used to create music and poetry. A creative agency used AI to generate music scores that were not only original but also resonated emotionally with listeners. By crafting prompts that set the mood and theme, the AI composed pieces that sounded as if they were penned by a human composer, blurring the lines between machine and human creativity. Similarly, in the realm of poetry, AI-generated verses have begun to make waves, offering fresh perspectives and styles that challenge our understanding of creativity itself. These innovative uses of prompt engineering show how AI can be more than a tool for efficiency—it can be a muse, inspiring new forms of art and expression.

The business impact of successful prompt engineering is hard to overstate. An e-commerce company leveraged AI to improve customer engagement through personalized recommendations. By crafting targeted prompts that considered customer behavior and preferences, they saw a significant uptick in sales and user satisfaction. It was like having a personal shopper for each customer, guiding them to products they didn't even know they wanted. The return on investment was clear: not only did they see increased revenues, but they also fostered a loyal customer base that kept coming back for more. These case studies illustrate how prompt engineering can turn AI from a novelty into a powerhouse of productivity and creativity. It's about crafting the right questions to unlock the full potential of AI, making it a valuable partner in achieving business objectives.

In wrapping up this chapter, we've explored how prompt engineering can transform AI from a mere tool into a dynamic ally across industries. From healthcare to creative arts, the ripple effects of well-crafted prompts are vast and varied. As we look ahead, the next chapter will delve into the exciting world of AI monetization, exploring how the skills you've honed can translate into financial opportunities. Stay tuned as we continue this exploration of AI's potential and power.

Chapter Six

Monetizing AI

Turning Skills into Income

YOU'RE ENJOYING YOUR MORNING coffee, the sky doing its best impression of a watercolor painting, when your phone buzzes with good news—another client, signed and sealed, while you were still in your slippers. There's something undeniably satisfying about making money while the kettle's still warm.

But that kind of freedom doesn't just happen by accident. It starts with spotting the right AI monetization opportunities—ones that actually fit your strengths, interests, and the way you want to live and work. Whether you're just getting curious about AI or you're a seasoned pro looking to shift gears, this landscape is wide open. And right now, it's begging for explorers.

First, let's explore how AI is making waves across various industries, creating opportunities you can tap into. In healthcare, AI is the digital superhero, revolutionizing diagnostics and patient management. Imagine AI-powered systems that analyze medical images with precision, aiding doctors in diagnosing conditions faster and more accurately. This not only saves time but also improves patient outcomes. Meanwhile, in finance, AI takes the role of a savvy analyst, optimizing trading systems and assessing risks with an accuracy that leaves human counterparts in awe. Automated trading systems execute trades at lightning speed, capitalizing on market fluctuations like a seasoned day trader who never sleeps.

Now, let's turn our attention to emerging markets where AI is poised to make a substantial impact. The renewable energy sector, for instance, is buzzing with potential. AI-driven smart grid management systems optimize energy distribution, reducing waste and enhancing efficiency. It's like having a conductor orchestrating an energy symphony, ensuring every watt is utilized to its fullest. In agriculture, AI is the modern farmer's best friend, enabling precision farming techniques. By analyzing soil health and weather patterns, AI helps farmers maximize crop yields and reduce resource usage. It's as if the fields themselves whisper secrets to the machines, ensuring bountiful harvests year after year.

But how do you identify which of these opportunities aligns with your skills and interests? Start by conducting a skills inventory. Take a good, hard look at what you're good at and where you might need to brush up. Are you a whiz with data, or do you have a knack for creative problem-solving? Understanding your strengths and weaknesses is like having a map before setting off on a journey. Next, engage in interest mapping to pinpoint what truly excites you. Whether it's healthcare, finance, or sustainable farming, aligning your monetization strategy with your passions ensures a fulfilling and rewarding venture.

Once you've identified your niche, it's time to evaluate market demand for AI services. This is where research becomes your best ally. Tools like Google Trends and industry reports can provide valuable insights into what services are hot and which ones are not. Imagine having a crystal ball that reveals market dynamics and customer preferences. By identifying service gaps, you can position yourself as a go-to expert, ready to address unmet needs. It's like finding a missing puzzle piece and knowing exactly where it fits. This strategic approach not only enhances your chances of success but also sets the stage for long-term growth and profitability.

Exercise: Mapping Your AI Journey

Grab a pen and paper, or open your favorite note-taking app. List your current skills and interests, then brainstorm potential AI-related opportunities that align with them. Use online tools to research market demand and identify service gaps. Reflect on how you can leverage your unique strengths to create value in this rapidly evolving landscape. This exercise will help you chart a course toward monetizing your AI skills effectively.

As you embark on this exciting journey, remember that the key to successful AI monetization lies in aligning your skills and interests with market demand. By exploring diverse industries, recognizing emerging trends, and strategically positioning yourself, you can unlock the full potential of AI and turn your skills into a profitable venture.

Freelancing with AI: Skills in Demand

Let's talk freelancing. It's like being your own boss without the hassle of firing yourself when you take an extra-long lunch break. In the AI world, freelancing offers a buffet of opportunities, especially if you've got the right skills in your toolkit. First up, there's machine learning model development. Imagine being the architect who designs the brain of a robot. You're not just building models; you're crafting the future of technology. These models can predict trends, automate tasks, or even suggest the next viral cat meme. It's a skill so hot, it's practically sizzling on the freelance market. Then there's natural language processing. Fancy term for teaching machines to understand human language. Whether it's for crafting chatbots or creating content, NLP is your ticket to making AI feel like a friendly companion rather than a cold machine.

Once you've polished those skills, it's time to hit the freelance platforms. Think of Upwork, Fiverr, and Freelancer.com as your digital marketplace. Here, clients are shopping for talent, and you're displaying your skills like a boutique with only the finest AI services. But if you're looking to swim in niche waters, Toptal is where the crème de la crème of freelancers hang out. These platforms are like bustling bazaars, where your profile is your storefront, and ratings are as precious as gold stars from your favorite teacher. Make your mark by showcasing your projects, client feedback, and a snazzy portfolio that screams, "I'm the AI expert you've been searching for!"

Speaking of portfolios, let's build one that could make even the most skeptical client say, "Take my money!" Your portfolio is your best friend, so fill it with past projects and case studies that showcase your prowess. Think of it as your digital brag book, where you highlight the skills and competencies that set you apart. Did you save a company thousands by automating their customer service? Or perhaps you developed a machine learning model that predicts trends faster than a teenager on TikTok? Let these stories shine like diamonds under a spotlight. Remember, a compelling portfolio is not just about what you've done but how you did it, and what you can do next.

Networking might sound like a chore, but it's more like a treasure hunt for opportunities. Start by leveraging LinkedIn. Consider it your digital Rolodex, but cooler and less likely to get lost under a pile of old magazines. Connect with professionals, join groups, and share your insights like the AI guru you are. Next, step out from behind the screen and attend industry conferences and meetups. These events are like speed dating for professionals, where you can rub shoulders with potential clients and collaborators. The key is to be genuine—people can spot a sales pitch a mile away. Instead, focus on building relationships and sharing your passion for AI, because sometimes, the best clients come from the most unexpected conversations.

Creating AI-Driven Products and Services

Imagine you're an inventor brimming with ideas, ready to create the next big thing in AI. But before you start building, it's crucial to identify market needs. Think of it like baking a cake; you don't want to whip up a chocolate masterpiece for a crowd craving pumpkin pie. Start by conducting customer surveys and focus groups. These tools are your magnifying glass, giving you insights into what people really want. By asking the right questions, you'll uncover unmet needs and pain points. It's like being a detective, piecing together clues to reveal what your AI product should deliver. Analyze competitor offerings to spot gaps in the market. Maybe there's a feature everyone seems to overlook, or a problem no one has solved yet. This is your chance to swoop in like a superhero and offer a solution that stands out in a crowded field.

Once you've nailed down the market needs, it's time to design AI solutions that address real-world problems. Picture AI-powered personal finance apps that help users manage their money like a pro. They could analyze spending habits, suggest saving strategies, and even remind users when bills are due. It's like having a financial advisor in your pocket, minus the hefty fees. Or consider AI-driven educational tools that offer personalized learning experiences. Imagine software that adapts to each student's pace, providing tailored lessons and exercises. It's like having a private tutor who knows exactly when to push and when to slow down. Designing these solutions requires creativity and a keen understanding of your audience's needs, ensuring your product isn't just cool, but truly valuable.

Now, let's talk about prototyping and testing, the sandbox phase where ideas take shape. Start by building a minimum viable product (MVP), a stripped-down version of your AI solution that includes only the core features. This MVP is your testing ground, allowing you to gather user feedback without investing too much time or money upfront. It's like test-driving a car before committing to the purchase. Gather feedback from real users and iterate based on their experiences. Maybe they love the app's functionality but find the interface clunky, or they really want additional features. This iterative process is your guide to refining the product, ensuring it meets user expectations and functions seamlessly.

Once you've polished your AI product, it's time to think about marketing and launch strategies. Digital marketing tactics are your megaphone, broadcasting your product to the world. Use targeted ads to reach your ideal audience, and create engaging content that showcases the benefits of your AI solution. Social media and online communities are your stage, offering platforms to engage directly with potential users. Build a community around your product, encourage user-generated content, and foster discussions. It's like hosting a party where everyone's talking about how great your product is. Leverage these platforms to spread the word and generate buzz, ensuring your AI-driven product gains the attention it deserves.

Visual Element: AI Product Development Checklist

Create a checklist to guide your AI product development process. Include tasks like identifying market needs, designing solutions, prototyping and testing, and planning marketing strategies. Use this checklist to track progress and ensure you're covering all bases as you bring your AI product to life. This visual guide will help you stay organized and focused, minimizing overlooked details and maximizing your product's success potential.

Building a Side Business with AI Innovations

Imagine you're sitting at your kitchen table on a lazy Sunday morning, sipping coffee and thinking about how to make those brilliant AI ideas of yours pay off. You're not alone. Many folks are diving into the world of AI side businesses, turning what once seemed like science fiction into profitable ventures. Let's start with brainstorming potential ideas. Picture this: AI-driven content creation services that whip up articles faster than you can say "writer's block." These platforms can generate everything from engaging blog posts to punchy social media updates, leaving you with more time to focus on strategy. Or how about a virtual assistant business powered by AI? These digital helpers can handle customer inquiries, schedule appointments, and even manage social media accounts, all while you kick back and enjoy your morning brew. It's like having a team of tireless employees without the water cooler gossip.

Once the ideas are flowing, it's time to put them into action with a solid business plan. This isn't just a dull document you toss in a drawer; it's your roadmap to success. Start by defining your business goals and objectives. What do you want to achieve with your AI venture? Maybe it's to revolutionize content creation or become the go-to virtual assistant service in your niche. Whatever it is, write it down. Next, tackle financial planning and budgeting. It's not the most exciting part, but it's crucial. Calculate your startup costs, expected revenue, and potential expenses. This financial blueprint helps you stay on track and avoid those "uh-oh" moments when you realize you've spent your marketing budget on fancy office chairs.

With your plan in hand, you're ready to secure funding and resources. Think of this stage as assembling your dream team, but with money. Options abound, from venture capital and angel investors to crowdfunding platforms like Kickstarter. Venture capital can provide a hefty boost if you're willing to part with some equity, while angel investors might be swayed by your passion and vision. Crowdfunding is another avenue, where you pitch your AI idea to the masses. It's like a modern-day bake sale, but instead of selling cookies, you're offering a slice of the future. You'll need to craft a compelling pitch and video that showcases your product's potential, sparking excitement and investment.

Once your business is off the ground, the next challenge is scaling and growing. It's like caring for a plant; you've got to nurture it to see it flourish. Focus on strategies for customer retention and acquisition. Offer top-notch service that keeps clients coming back, and use targeted marketing to reach new audiences. Building strong relationships with your customers is key. Explore partnerships and collaborations that can propel your business to new heights. Maybe there's a tech company that could benefit from your AI expertise, or a marketing firm looking for cutting-edge content solutions. These partnerships can open doors to new markets and opportunities, accelerating your growth.

So there you have it, the blueprint for building a thriving AI side business. From brainstorming innovative ideas to crafting a detailed business plan, securing funding, and scaling your operations, each step brings you closer to transforming your AI dreams into reality. Whether it's revolutionizing content creation or streamlining business processes with virtual assistants, the possibilities are as vast as your imagination. With dedication and a sprinkle of creativity, you can turn those Sunday morning musings into a thriving venture that not only fills your wallet but also brings a sense of fulfillment and achievement.

Prompts for Building a Side Business with AI Innovations

Brainstorming AI-Powered Business Ideas

"Act as an AI business idea generator. Suggest five AI-driven side business ideas that could be started with minimal upfront investment and can be operated remotely."

"Pretend you are a creativity coach. Help me brainstorm AI-based services I could offer to small businesses, such as virtual assistant support, content generation, or customer service automation."

Building a Solid Business Plan

"Imagine you are a business planning mentor. Guide me through creating a simple AI side business plan, including goals, target market, services offered, and basic financial planning."

"Act as a startup strategist. Help me map out a one-page business plan for launching an AI-powered content creation service for niche blogs."

Funding and Resource Strategies

"Pretend you are a funding advisor. Suggest beginner-friendly ways I could raise money or bootstrap my AI side business without needing venture capital."

"Imagine you are a crowdfunding coach. Walk me through how to create a compelling Kickstarter pitch for an AI-driven productivity app designed for freelancers."

Scaling and Growing the Business

"Act as a business growth consultant. Suggest three strategies for acquiring and retaining customers for an AI-based virtual assistant service."

"Pretend you are a partnership strategist. Help me identify potential collaboration opportunities that could help expand my AI business into new markets."

Success Stories: Entrepreneurs Leveraging AI

Let me share some tales of trailblazers who have turned AI into a goldmine, starting with the healthcare sector. Picture a bustling startup named MedTech Innovators. They focus on AI diagnostics, where machine learning models scan medical images to detect anomalies faster than a seasoned doctor. Imagine the relief for healthcare professionals who can now focus more on patient care and less on deciphering complex images. This San Francisco-based startup didn't just stop at diagnostics; they ventured into patient management systems powered by AI, streamlining processes and reducing administrative burdens. The result? A 40% increase in efficiency and a significant reduction in healthcare costs. It's like giving healthcare a turbo boost, all thanks to AI.

Over in the financial realm, FinNext Solutions has made waves by revolutionizing automated trading systems. These folks created an AI-driven platform that analyzes market trends in real-time, executing trades with precision that would make even the savviest trader envious. By leveraging AI algorithms, they've managed to increase portfolio returns while minimizing risks. The secret sauce? Their AI continuously learns from market data, adapting its strategies to stay ahead of the curve. It's like having a financial advisor who never sleeps, ensuring your investments are always in safe hands. This level of adaptability and innovation is what sets successful AI entrepreneurs apart, allowing them to navigate challenges in AI development with finesse.

Now, let's talk about the magic happening in e-commerce. Think of a dynamic e-commerce platform called ShopSmart. It uses AI to enhance customer experiences through personalized shopping recommendations.

Imagine logging in to find a curated selection of items that feel handpicked just for you. This level of personalization not only boosts customer satisfaction but also drives sales, turning casual browsers into loyal buyers. ShopSmart's AI analyzes browsing patterns and purchase histories, creating a seamless shopping experience akin to having a personal shopper at your service. It's these kinds of AI-driven innovations that transform how businesses engage with customers, proving that a little tech-savvy can go a long way in e-commerce success.

But AI entrepreneurship isn't just about profits; it's also about making a difference. Enter GreenTech Innovations, a startup dedicated to environmental sustainability. They're harnessing AI to tackle climate change, deploying AI models to optimize energy consumption and reduce waste. Picture smart grids that adjust energy usage based on demand, cutting down on unnecessary consumption and lowering carbon footprints. It's like giving the planet a much-needed breather, one kilowatt at a time. These initiatives are not only inspiring but also highlight the role of AI in driving social impact. By developing AI solutions that address environmental challenges, these entrepreneurs are paving the way for a greener, more sustainable future.

Looking ahead, the future of AI entrepreneurship is brighter than ever. One emerging trend is the use of AI in cybersecurity advancements. As cyber threats become more sophisticated, AI steps in as a digital bodyguard, detecting and neutralizing threats before they cause harm. Entrepreneurs who tap into this field are poised to make significant contributions to data protection and privacy. Another exciting frontier is AI applications in remote work and collaboration tools. Imagine AI systems that streamline communication, manage projects, and even predict team productivity. As remote work becomes the norm, these tools will be indispensable, offering entrepreneurs a chance to shape the future of work.

In wrapping up this chapter, we've seen how AI is not just reshaping industries but also redefining possibilities for entrepreneurs. From healthcare innovations to e-commerce transformations, these success stories highlight the transformative power of AI. As we continue to explore the potential of AI, remember that the opportunities are as boundless as your imagination and that the next big success story could be yours.

Chapter Seven

AI and Career Advancement

Future-Proofing Your Path

Let's say you step into a job fair where the resumes are digital, the interviews are algorithm-assisted, and career advice comes with a side of machine learning. No, it's not a scene from the next sci-fi blockbuster—it's the very real, rapidly evolving landscape of modern work.

AI has moved from the realm of techy daydreams to an active, behind-the-scenes force reshaping industries, roles, and opportunities. It's streamlining repetitive tasks, surfacing strategic insights, and giving professionals across the board a powerful new ally. Whether you're climbing the corporate ladder, pivoting careers, or launching your own venture, AI has quietly taken a seat at the table—and it's ready to help you go further, faster.

AI is transforming careers across the board, much like a master chef adding spices to a bland dish. It introduces flavors and nuances that weren't there before, spicing up job roles and responsibilities in unexpected ways. Imagine moving from a traditional IT role to one that focuses on AI project management and strategy. This isn't just a shift in job title; it's a leap into a realm where creativity meets technology. AI is creating new job categories, opening doors to positions in AI development and management that didn't exist a decade ago. It's like discovering a secret passageway in a mansion you've lived in for years, and suddenly, there are new rooms filled with opportunities.

For those wondering where AI opportunities might hide within their current fields, the answer could be as close as your next project. AI project management roles are cropping up like wildflowers, offering chances to lead teams in implementing cutting-edge technologies. Whether you're in healthcare, finance, or even agriculture, AI research and development roles are waiting to be explored. Think of it as adding a new tool to your belt, one that makes you indispensable in an AI-driven workplace. It's not about discarding your current skills but enhancing them with AI knowledge, like upgrading from a trusty old toolbox to a state-of-the-art gadget kit.

Considering a career transition but unsure how to make the leap? AI skills might just be your golden ticket. Transitioning from traditional roles to AI-specialized positions is akin to trading in a bicycle for a sports car. You gain speed, efficiency, and a whole new perspective on the road ahead. AI certifications can help pivot your career into tech fields, offering the credentials needed to stand out in a competitive job market. It's like having a passport that opens doors to new countries, except in this case, those countries are filled with exciting career possibilities.

But AI isn't just about new roles; it's also about enhancing your current job performance. Imagine having a personal assistant that handles mundane tasks while you focus on strategic initiatives. AI tools for decision-making and analytics are like having a wise old sage whispering advice into your ear, offering insights that help you navigate even the trickiest of situations. By automating repetitive tasks, AI frees up your time, allowing you to focus on what truly matters—like brainstorming the next big idea or finally getting around to organizing that office drawer.

Interactive Element: AI Career Mapping Exercise

Grab a notebook and sketch out your current job role. Identify areas where AI could integrate or enhance your responsibilities. Consider potential AI-related roles you could transition into and list any skills or certifications needed. Reflect on how AI fits into your career path and how you can leverage it for growth.

As you explore the transformative potential of AI in your career, you'll find that it's not just a tool but a partner in your professional journey. Whether you're looking to climb the corporate ladder or pivot into a new field, AI offers the resources and insights needed to navigate the ever-evolving world of work.

Skills for the AI-Driven Job Market

Imagine being at a potluck where everyone's dish represents a different skill, and you're tasked with bringing the secret sauce that brings it all together. In the AI-driven job market, that secret sauce comes in the form of specific technical skills. Proficiency in programming languages like Python and R is as essential as salt in the kitchen. Python is the go-to language for AI, thanks to its simplicity and vast library support, which makes it perfect for developing machine learning algorithms. Meanwhile, R is the data scientist's best friend, especially when it comes to statistical analysis and data visualization. Understanding machine learning algorithms is like knowing how to drive a stick shift—it opens up a world of possibilities. Data analysis skills are crucial, too. They allow you to make sense of the mountains of data AI systems generate, turning numbers into actionable insights. It's like being a detective, sifting through clues to unravel the mysteries hidden in data sets.

Now, let's not forget the softer side of skills. Imagine AI as a dance partner—it's not just about knowing the steps but also about moving in harmony with others. Critical thinking and problem-solving become your choreography, enabling you to adapt and respond to AI-generated insights. It's about looking beyond the numbers and asking, "What does this mean?" and "How can I use this to solve real-world problems?" Effective communication is the rhythm that keeps the team in sync. In AI projects, where technical jargon can often sound like a new language, being able to translate complex concepts into simple terms is invaluable. Picture yourself in a meeting, explaining an AI model to a group of non-tech people—nailing that explanation is like hitting the right notes in a song, ensuring everyone's on the same page.

But here's the kicker: the AI world never stands still. It's like a never-ending rollercoaster, where the track is constantly being extended. Continuous skill development is your safety harness, keeping you from flying off the ride. Participating in AI workshops and bootcamps is a fantastic way to stay updated. These events immerse you in the latest tech, akin to a crash course in the newest dance moves. Online platforms like Coursera and edX offer a smorgasbord of courses. Whether you're a newbie or a seasoned pro, there's always something new to learn. These platforms are like a treasure trove of knowledge, waiting to be explored at your own pace.

And then there's the cherry on top: certifications and qualifications. They're like the scout badges of the AI world, showcasing your skills and dedication. Certifications such as the Google Cloud AI Engineer certification or the Microsoft Azure AI Fundamentals certification not only enhance your resume but also give you a competitive edge. They signal to employers that you're committed to staying at the forefront of the industry. It's like wearing a badge of honor that says, "I know my stuff, and I'm ready to take on any AI challenge." These qualifications can open doors to new opportunities, acting as a key to unlock roles that require a higher level of expertise.

In this ever-evolving field, staying ahead means being proactive. It's about mixing the right ingredients—technical skills, soft skills, continuous learning, and certifications—to create a recipe for success.

Integrating AI into Your Professional Skillset

Imagine you're in the world of marketing, where understanding your audience feels like trying to catch a fish with your bare hands. Enter AI, your trusty net, ready to scoop up insights and make sense of the chaos. In marketing, AI can be your best friend for customer segmentation and targeting. It's like having a spotlight that illuminates the specific preferences and behaviors of your audience, allowing you to tailor campaigns with laser-like precision. Instead of casting a wide net and hoping for the best, AI helps you focus on the individuals who are most likely to engage with your brand. Predictive analytics in supply chain management is another game-changer. Imagine having a crystal ball that anticipates demand, optimizing inventory levels and ensuring that you're never caught off guard by unexpected shortages or surpluses. AI brings this kind of foresight to the table, transforming supply chains into well-oiled machines that respond dynamically to changing conditions. By integrating AI applications into your specific professional skillset, you can enhance decision-making and drive efficiency like never before.

Now let's talk about getting cozy with AI tools and software. Picture yourself as a kid in a candy store, except the sweets are AI tools like TensorFlow and IBM Watson. TensorFlow is your go-to for machine learning projects, offering a flexible framework to build models that can analyze data, recognize patterns, and make decisions with minimal human intervention. It's like having a personal assistant who learns from experience, getting smarter and more efficient with every task. On the other hand, IBM Watson is your partner in implementing AI models that can understand and interact with humans in natural language. Think of it as your digital translator, capable of turning complex data into easy-to-understand insights. By learning these tools, you're not just adding some extra tricks to your repertoire; you're equipping yourself with powerful allies that can elevate your professional game.

But theory only takes you so far—it's time to roll up those sleeves and dive into practical AI projects. Start by developing a chatbot for customer service inquiries. Imagine creating a digital concierge that can handle routine questions, freeing up your human team to focus on complex issues. It's like having a helpful sidekick who never tires and always has the answers. Or consider building a predictive model for sales forecasting. This project lets you analyze historical data to predict future sales trends, offering insights that can inform strategic decisions. Picture yourself as a fortune teller for your business, with AI as your crystal ball, providing clarity in an ever-changing market. Engaging in practical projects like these not only deepens your understanding of AI but also showcases your ability to apply knowledge in real-world scenarios.

Once you've honed your AI skills, it's time to show them off in your professional profile. Think of your resume as a stage, and your AI projects as the star performers ready to take the spotlight. Create a dedicated section for AI projects, highlighting the skills and technologies you've mastered. Use quantifiable metrics to demonstrate the impact of your work. Instead of saying, "Built a chatbot," try "Developed a chatbot that reduced customer response time by 40%." Numbers speak louder than words, offering concrete evidence of your achievements. By effectively showcasing your AI skills, you position yourself as a valuable asset in a tech-driven world, ready to tackle challenges and drive innovation.

Prompts for Integrating AI into Your Professional Skillset

Applying AI to Your Industry

"Act as a career integration coach. Suggest three ways I could apply AI tools in my current profession (marketing, supply chain management, customer service, etc.) to improve efficiency and decision-making."

"Pretend you are an AI application strategist. Help me brainstorm a practical AI project that fits my field and could be completed within 30–60 days to boost my skills."

Learning Key AI Tools and Software

"Imagine you are an AI tool tour guide. Recommend a beginner-friendly learning path for mastering essential AI tools like TensorFlow or IBM Watson, based on my interest in customer insights and automation."

"Act as an AI skills advisor. Suggest which AI tools would be most useful for someone working in [insert profession: marketing, logistics, healthcare, education, etc.] , and why."

Hands-On AI Projects

"Pretend you are a project coach. Guide me through a basic project to build a customer service chatbot using free or low-cost AI tools."

"Imagine you are an AI project mentor. Help me set up a predictive analytics project to forecast sales, inventory needs, or customer engagement using historical data."

Showcasing AI Skills in Professional Profiles

> "Act as a resume upgrade consultant. Help me write a powerful bullet point for my LinkedIn profile highlighting an AI project I completed (like developing a chatbot or a predictive model)."

> "Pretend you are a portfolio strategist. Suggest how I could structure an online portfolio that showcases my AI projects and attracts potential employers or clients."

Networking and Building an AI Community

Imagine you're at a neighborhood block party, but instead of burgers and lemonade, you're surrounded by AI enthusiasts, data scientists, and tech innovators. This is the world of AI networking, where forming connections can be as vital as learning to code. Networking in AI isn't just about attending events and collecting business cards. It's about building meaningful relationships with industry professionals who can offer insights, opportunities, and maybe even a job lead or two. Think of platforms like LinkedIn as your digital Rolodex, where you can connect with individuals who share your passion for AI. By joining AI-focused professional organizations, you're not only expanding your network but also gaining access to exclusive events, resources, and mentorship opportunities that can propel your career forward.

Attending AI conferences and events is like being at a concert where instead of music, innovation is the star of the show. Events like AI Expo are not just about showcasing the latest technologies; they're about engaging with the minds behind those breakthroughs. These gatherings are fertile ground for exchanging ideas, learning from experts, and spotting emerging trends. Imagine the buzz at NeurIPS or AI Summit, where the air is filled with discussions on machine learning advancements, ethical AI, and the future of automation. It's where you can rub elbows with thought leaders and innovators, perhaps over a cup of overpriced conference coffee, but the connections you make are worth every sip. Such events are not only informative but also provide networking opportunities that can lead to collaborations, partnerships, or even your next career move.

The internet is a vast ocean, and within it are vibrant AI online communities where enthusiasts, novices, and experts converge. Engaging in platforms like Reddit or Stack Overflow is akin to joining an ongoing conversation where everyone's invited. These forums are treasure troves of knowledge, where you can contribute to discussions, seek advice, or simply learn from the experiences of others. Joining AI-focused groups on social media platforms allows you to stay updated with the latest news, trends, and debates in the AI world. It's like having a front-row seat to a never-ending conference, accessible anytime from your favorite armchair. Participating actively can enhance your understanding of complex topics, introduce you to diverse perspectives, and help you build a reputation as a knowledgeable and engaged member of the AI community.

Mentorship and collaboration are the secret sauce to thriving in the AI landscape. Finding a mentor who's an experienced AI practitioner can provide guidance, support, and insider tips on navigating the industry. Imagine the value of having someone who's been through the trenches, sharing their wisdom and helping you avoid common pitfalls. It's like having a personal GPS for your career. Additionally, collaborating on open-source AI projects is a fantastic way to gain practical experience, showcase your skills, and contribute to the community. These projects are like collaborative art pieces, where each contributor adds their unique touch, resulting in a masterpiece that benefits everyone. By working together, you not only enhance your skills but also demonstrate your ability to collaborate effectively, a trait highly valued in the tech world.

Staying Relevant: Continuous Learning in AI

In the fast-paced world of artificial intelligence, trying to keep up can feel like racing against a speeding train while balancing a cup of coffee in one hand and a donut in the other. But staying relevant in AI is crucial, and it all starts with embracing continuous education. Technological advancements are happening at lightning speed—blink, and you might miss the next big thing. This isn't just about keeping your skills from gathering dust; it's about adopting a growth mindset that propels both your personal and professional development. Think of it as a never-ending adventure where each new skill learned or concept grasped is another feather in your cap. This mindset encourages you to view challenges as opportunities rather than obstacles, turning every hiccup into a stepping stone toward mastery.

Luckily, there are plenty of resources to fuel your learning journey. Online platforms like Udacity's AI Nanodegree offer structured courses designed to bring you up to speed with the latest AI techniques. They're like having a personal tutor on standby, ready to guide you through the complex maze of data science and machine learning. If you're the type who enjoys curling up with a good read, AI research papers and journals can be your best friends. They provide cutting-edge insights into the field, keeping you informed about groundbreaking research and emerging trends. It's like having a front-row seat to the latest scientific breakthroughs, minus the lab coat. These resources serve as a treasure trove of information, empowering you to expand your knowledge base and sharpen your skills.

Maintaining competency in AI is like staying in shape; it requires regular workouts to keep your skills toned and ready for action. Setting learning goals and milestones is a great way to stay on track. Whether it's mastering a new programming language or understanding a complex algorithm, having clear objectives ensures progress. Allocate time each week for AI study sessions, treating them like appointments you can't miss. This consistency builds a habit, transforming learning from a chore into a rewarding routine. It's like planting seeds in a garden; with regular care, they grow into something beautiful and productive. This approach not only keeps your skills sharp but also fosters a sense of accomplishment as you tick off each milestone.

Evaluating and adapting to new AI trends is akin to surfing on the crest of a technological wave. Industry reports are invaluable for spotting these trends, offering a bird's-eye view of where the field is headed. They help you anticipate changes and position yourself strategically, much like a chess player thinking several moves ahead. Experimenting with new AI tools and platforms is another way to stay ahead of the curve. It's like being a kid in a tech playground, where trying out different gadgets and tools keeps your skills fresh and adaptable. These experiments can lead to unexpected insights or even spark ideas for innovative projects. By staying curious and open to new experiences, you ensure that you're not just keeping up but actively shaping the future of AI.

As you continue to explore the ever-expanding universe of AI, remember that staying relevant is a journey, not a destination. Embrace the challenges, celebrate the small wins, and keep your eyes on the horizon. Your willingness to learn and adapt will serve as your compass, guiding you through the exciting landscape of artificial intelligence. In the next chapter, we explore how AI is revolutionizing creative industries, opening doors to new realms of innovation and artistry.

Chapter Eight

Ethical AI

Responsible and Conscious Use

DEVELOPING ETHICAL AI IS a bit like juggling flaming torches while riding a unicycle through a crowded carnival. It's thrilling, full of potential, and just one misstep away from chaos. One moment, you're dazzling the crowd; the next, you're wondering if someone remembered to call the fire department.

Building AI responsibly isn't just about writing clever code or crafting sleek algorithms—it's about finding the right balance. Like baking a temperamental soufflé, you need the right mix of fairness, accountability, and transparency or the whole thing could fall flat (or worse, explode spectacularly). Because while AI can unlock amazing possibilities, it can just as easily veer into "Black Mirror" territory if we're not careful. And no one wants to be the one who programmed the toaster to become self-aware.

At its core, ethical AI is about doing the right thing, even when no one's watching. It's about making sure our digital companions behave more like well-mannered guests at a dinner party rather than unruly toddlers on a sugar high. This means designing AI systems that are fair, transparent, and accountable. Tech giants like Google and Microsoft have laid down ethical frameworks to steer their AI creations towards the path of righteousness. Google, for instance, employs a federated approach, ensuring that every layer of their organization is imbued with ethical considerations. Meanwhile, Microsoft takes a top-down approach, with oversight from a Responsible AI Council and an AI Ethics Committee. These frameworks act like guardrails on a rollercoaster, keeping the ride thrilling yet safe.

The principles guiding ethical AI are akin to the moral compass of a superhero—always pointing towards justice and equality. Respect for human rights and dignity takes the front seat, ensuring AI systems don't go rogue and start making decisions that could negatively affect individuals or communities. Inclusivity and diversity in AI development are crucial, ensuring that our AI doesn't only cater to one type of user but embraces the kaleidoscope of humanity. It's like hosting a dinner party where everyone gets a seat at the table and a dish that suits their taste. This approach ensures that the AI we create reflects the richness and diversity of our world, rather than a narrow slice of it.

In practice, ethical AI is about more than just good intentions; it's about real-world implementation. Take healthcare, for example, where AI can ensure unbiased patient treatment by analyzing medical data without preconceived notions. This means a patient's care is based on data-driven insights rather than stereotypes or assumptions. In education, AI tools maintain equitable access by tailoring learning experiences to individual needs, much like a personal tutor who understands your quirks and learning style. These examples show that ethical AI isn't a pipe dream; it's a reality that's reshaping industries for the better.

However, implementing ethical AI isn't always a walk in the park. It's more like navigating a minefield while blindfolded. One of the biggest challenges is balancing profit with ethical responsibilities. Companies might be tempted to cut corners for the sake of the bottom line, but doing so risks turning AI into an unpredictable monster rather than a helpful ally. Overcoming resistance to ethical guidelines in corporate culture is another hurdle. Some organizations might view ethics as a roadblock to innovation, akin to a speed bump on the highway to success. But embracing ethical AI is more like installing airbags in a car; it's a safeguard that ensures long-term sustainability and trustworthiness.

Interactive Element: Ethical AI Reflection Exercise

Take a moment to consider a recent AI tool or application you've encountered. Reflect on the following questions:

- Was the AI fair and unbiased in its operation?
- Did it provide transparency in how decisions were made?
- How well did it respect individual rights and privacy?

Jot down your thoughts and consider how these reflections might influence your approach to engaging with AI technologies in the future.

By embracing ethical AI principles and practices, we can ensure that our technological innovations serve humanity with integrity and foresight.

Navigating AI Bias: Challenges and Solutions

Picture AI as your trusty yet occasionally misguided sidekick. It's great for many things, but sometimes it stumbles into biases that are as sneaky as the last cookie in the jar. AI bias happens when these systems make decisions that unfairly favor or disadvantage certain groups. Think of it as your GPS leading you into a lake because it can't tell the difference between a road and a boat ramp. This bias can manifest in various ways, such as facial recognition technology that struggles to accurately identify people of color, women, and nonbinary individuals. It's like having a camera that only takes good photos when the lighting is just right. Similarly, hiring algorithms might unintentionally favor certain demographics over others, leading to workplace discrimination. Imagine the algorithm as a gatekeeper who only lets certain people into the club based on a quirky, outdated dress code.

The roots of AI bias can often be traced back to the data used to train these systems. If the training data is skewed, the AI will be too. It's like teaching a child math using a textbook with half the pages missing. The lack of diversity in development teams also contributes to bias. When the people designing AI systems don't reflect the diversity of the world, the systems they create might fail to account for different perspectives and experiences. Picture a team of chefs trying to create a menu for a global restaurant chain based solely on their hometown cuisine. The result might be delicious, but it won't cater to a variety of tastes.

To tackle AI bias, we need a game plan as strategic as a chess master's opening move. One effective strategy is to use diverse and representative datasets during the training phase. Think of it as expanding the playlist at a party to include everyone's favorite tunes, ensuring no one feels left out. Regular audits and bias detection tools are also essential. They act like quality control inspectors, checking for signs of bias and helping to correct any missteps. It's about keeping the system honest and ensuring it plays fair, much like a referee at a soccer match.

Case studies provide a glimpse into how bias has been successfully identified and mitigated. For example, in law enforcement, AI tools have undergone revisions to eliminate bias in facial recognition. These changes ensure that everyone is treated fairly, regardless of their appearance. In the world of recruitment, AI-driven platforms have been improved to prevent bias in hiring decisions. These platforms now focus on skills and qualifications rather than irrelevant factors, ensuring that the best candidate gets the job. It's like having a talent scout who evaluates musicians based solely on their performance, not their stage attire.

Case Study: Bias in AI Recruitment

Consider an AI system initially biased against certain universities because its training data favored graduates from a handful of prestigious schools.

By expanding the dataset to include a wider range of institutions and focusing on skills rather than school names, the bias was reduced. This change not only enhanced diversity but also uncovered talent from unexpected places, showing that sometimes the best gems are found off the beaten path.

The journey to eliminate AI bias is ongoing, and while it's a complex challenge, it's one worth tackling. Each step toward fairness and inclusivity ensures that AI serves everyone better, making it not just a technological marvel but a tool for positive change.

Privacy Concerns in AI: What You Need to Know

Imagine having a nosy neighbor who peeks through your windows, jotting down every little thing you do. That's what it can feel like when AI systems collect personal data without proper privacy protections in place. In today's digital age, privacy concerns are more pressing than ever, especially with AI's increasing role in collecting and using personal data. Data breaches are a prime example of the chaos that can ensue when privacy isn't prioritized. Just picture a treasure chest left open for all to see, with sensitive information spilling out for anyone to grab. These breaches can lead to identity theft, financial loss, and reputational damage. It's like leaving your front door wide open and inviting trouble in for tea. The role of AI in personal data collection is vast, ranging from tracking your shopping habits to analyzing your health data. While this can lead to personalized experiences and improved services, it also raises significant privacy concerns. You wouldn't want your diary read by strangers, so why let your digital footprint be any different?

Regulations like the General Data Protection Regulation (GDPR) have become the privacy police, setting boundaries for how data should be handled. The GDPR impacts AI data handling by enforcing strict guidelines on data collection, processing, and storage. It's like having a rulebook that ensures data is treated with the respect it deserves. Under GDPR, organizations must obtain clear consent from individuals before collecting their data, ensuring transparency and accountability. This regulation also mandates data minimization, meaning only the necessary data should be collected for a specific purpose. It's akin to packing for a trip and only taking what you need, rather than your entire wardrobe. Industry-specific privacy legislation further shapes how AI applications operate, ensuring that sectors like healthcare and finance adhere to rigorous standards. These laws protect sensitive information, acting as a shield against misuse and exploitation.

To tackle privacy concerns in AI, several techniques can be employed to protect personal data. Data anonymization and encryption serve as digital cloaks, disguising information so that it remains secure even if accessed by unauthorized parties. Think of it as wrapping your data in an invisibility cloak, making it unreadable to prying eyes. User consent frameworks play a crucial role in ensuring transparency. They provide individuals with clear, accessible information about how their data will be used, allowing them to make informed decisions. It's like reading the menu before ordering in a restaurant, knowing exactly what you're getting into. Transparency tools further enhance privacy by offering insights into data handling practices, building trust between organizations and users.

Balancing user privacy with AI system performance is like walking a tightrope—one misstep can lead to a tumble. There's often a trade-off between data usage and AI accuracy. More data can improve AI's performance, but it also increases privacy risks. It's like wanting to bake the perfect cake but not wanting to use too much sugar. Innovations in privacy-preserving AI technologies are emerging to address this challenge, offering solutions that maintain privacy without sacrificing performance. Techniques such as differential privacy and federated learning allow AI systems to learn from data without accessing it directly, making it possible to have your cake and eat it too. By employing these methods, organizations can create AI systems that respect privacy while delivering accurate and reliable results.

Prompts for Assessing and Implementing Privacy Protections in AI

Assessing Privacy Risks and Compliance

> "Act as a privacy risk assessor. Walk me through how to evaluate an AI system for potential privacy vulnerabilities, such as unnecessary data collection or lack of user consent."

> "Pretend you are a GDPR compliance coach. Help me create a basic checklist to ensure my AI project aligns with GDPR requirements for data collection, storage, and transparency."

> "Imagine you are a data protection auditor. Suggest three questions I should ask before launching an AI application that collects personal user data."

Implementing Privacy Protection Techniques

"Act as a data security advisor. Show me how to implement basic data anonymization and encryption methods to protect sensitive user information in an AI project."

"Pretend you are a user consent architect. Help me design a simple, clear consent framework that explains how user data will be collected and used before they agree."

"Imagine you are a transparency tool consultant. Suggest a beginner-friendly way to provide users with access to see what data an AI system collects about them and how it's used."

Balancing Privacy with AI Performance

"Act as a privacy-accuracy strategist. Help me decide when to prioritize stricter data privacy techniques (like differential privacy) even if it might slightly reduce my AI's performance."

"Pretend you are a federated learning coach. Explain how federated learning can help me train an AI model without collecting raw user data — and suggest a simple use case where I could try it."

Ongoing Monitoring and Improvement

"Imagine you are a privacy maintenance expert. Show me how to set up regular audits and updates to make sure my AI system stays compliant with evolving privacy laws and best practices."

Establishing Responsible AI Guidelines for Businesses

Creating a robust AI ethics policy is like setting the ground rules for a family game night—everyone needs to know what's expected to keep things fun and fair. For businesses looking to integrate AI into their operations, developing a clear and comprehensive ethics policy acts as a guiding star, ensuring that AI technologies align with the company's values and goals. Start by weaving ethics into the very fabric of your corporate strategy. Consider it as adding a dash of mindfulness to your tech endeavors. In practice, this means defining what ethical AI looks like for your organization and setting out clear guidelines for its development and deployment. Successful companies often use case studies to illustrate these guidelines, showing real-world examples of how their policies have been implemented effectively. The result is a playbook that not only safeguards against potential missteps but also fosters innovation and trust among stakeholders.

Now, once you've got your ethics policy drafted, the next step is to ensure that your team understands it inside and out. This is where AI ethics training comes into play. Think of it as the user manual for your ethics policy, filled with practical examples and engaging scenarios. Designing an effective training program requires a bit of creativity. You don't want your sessions to feel like a dreary lecture from your high school days. Instead, aim for interactive workshops that use case studies to bring the material to life. Imagine role-playing exercises where employees navigate ethical dilemmas, or group discussions that explore the nuances of AI ethics in different business contexts. These training programs should aim to equip your team with the knowledge and confidence needed to make ethical decisions in their day-to-day work.

Implementing an ethics policy isn't just about writing it down and hoping for the best. You need mechanisms to monitor and enforce these guidelines, much like a diligent gardener tending to their crops. This involves setting up AI ethics oversight committees that act as the custodians of your policy, ensuring it's followed and updated as needed. Think of them as the vigilant referees of your AI game, ensuring everyone plays by the rules. Regular audits and compliance checks are also crucial. They provide a snapshot of how well your organization is adhering to its ethical commitments and help identify areas for improvement. Consider these audits as routine health check-ups for your AI systems, keeping them in tip-top shape and aligned with your ethical standards.

The benefits of adopting ethical AI practices are manifold, much like the rewards of eating your vegetables. For starters, they enhance your brand's reputation and build consumer trust. Customers are more likely to engage with companies that demonstrate a commitment to ethical practices, seeing them as responsible and forward-thinking. This trust translates into loyalty, which in turn can drive growth and profitability. Ethical AI practices also help reduce legal risks and liabilities. By ensuring compliance with regulations and creating transparent AI systems, your business can avoid costly legal battles and reputational damage. It's like having a sturdy umbrella on a stormy day—providing protection when you need it most.

Balancing Innovation and Ethics: A Path Forward

Imagine you're a chef in a bustling kitchen, whipping up a new dish that could become your signature masterpiece. But there's a catch—you must balance bold flavors without overwhelming the palate. That's where the innovation-ethics dilemma comes into play in AI development. Rapid innovation can be like adding too much hot sauce; it might give a quick kick but could leave an unpleasant aftertaste if not handled with care. The tech world is buzzing with tales of AI breakthroughs, yet these advancements often come with ethical conundrums. For instance, developing facial recognition technology that can identify individuals in a crowd is a marvel of innovation. Still, it raises questions of privacy and potential misuse. Similarly, AI systems designed to optimize hiring processes can inadvertently perpetuate bias if not carefully calibrated. Balancing the competitive edge AI offers with ethical integrity is like tightrope walking—one misstep, and the consequences can be significant.

In encouraging ethical innovation, collaboration becomes the secret ingredient. Picture a bustling workshop where ethicists and technologists brainstorm together, each bringing unique perspectives to the table. This collaborative approach ensures that AI systems are not only cutting-edge but also aligned with ethical standards. Encouraging this synergy can be as simple as organizing cross-disciplinary teams to tackle AI projects, combining technical expertise with ethical insights. Incentives for ethical AI research and development serve as the cherry on top, motivating teams to prioritize ethics alongside innovation. Companies can offer awards, recognition, or even financial support for projects that demonstrate a commitment to ethical AI. This approach not only fosters a culture of responsibility but also positions ethical innovation as a competitive advantage in the marketplace.

Meanwhile, policymakers play a crucial role in this balancing act. Legislative frameworks are like the rulebook every player must follow, ensuring fair play in the AI arena. By establishing clear guidelines on what is permissible, policymakers provide a safety net that allows innovation to flourish within ethical boundaries. Public-private partnerships further strengthen this framework, creating a dialogue between governments and tech companies to develop comprehensive AI governance. Such collaborations ensure that policies remain relevant and adaptable to the fast-paced nature of technological advancements, much like a well-oiled machine that adapts to changing terrain.

Looking ahead, the future of ethical AI is brimming with potential and promise. Self-regulation within the AI industry is emerging as a proactive approach to uphold ethical standards. Companies are taking the initiative to establish their own guidelines, setting examples for others to follow. Industry standards offer a roadmap for ethical AI deployment, fostering consistency and trust across the board. At the same time, innovations in ethical AI monitoring tools provide an added layer of accountability. These tools act as watchdogs, scrutinizing AI systems for ethical compliance and flagging potential issues before they escalate. It's like having a safety inspector always on duty, ensuring that ethical considerations are woven into the fabric of AI development.

As we wrap up our exploration of ethical AI, it's clear that the path forward is one of balance and foresight. Ethical considerations should not be seen as obstacles but as guiding principles that enhance the value and impact of AI technologies. By fostering a culture that prioritizes ethics alongside innovation, we can create AI systems that not only push the boundaries of what is possible but also uphold the values we hold dear. Through collaboration, robust policies, and forward-thinking strategies, we can navigate the complexities of ethical AI, ensuring that our digital companions serve us with integrity and respect.

Chapter Nine

AI in Creative Industries

Unleashing Innovation

You walk into an art gallery and something feels... different. The paintings shimmer with unexpected energy, each one pulsing with a kind of imaginative spark that feels both familiar and otherworldly. But here's the twist—some of these works weren't crafted by human hands at all.

Welcome to the dazzling frontier of AI-generated art, where algorithms slip into metaphorical berets and go wild on digital canvases. Tools like DeepArt and Artbreeder are turning ordinary photos into fantastical masterpieces, often with nothing more than a few clicks and a bit of creative curiosity. It's like having a digital Picasso on standby—no messy paint tubes required.

DeepArt acts like a skilled artist-in-residence, taking your photo and transforming it into various art styles, from abstract to surreal. Meanwhile, Artbreeder allows you to blend and manipulate images, creating something entirely new from the familiar. It's like a digital palette where you're not just painting but also remixing reality itself. And then there are Generative Adversarial Networks, or GANs, which consist of two neural networks duking it out to create images so convincing, they could fool even the most seasoned art critic. GANs work by having one network generate images while the other acts as a discerning critic, refining them until perfection is achieved.

Yet, AI isn't just working solo; it's teaming up with artists to create hybrid masterpieces that blur the lines between man and machine. Imagine an artist collaborating with AI to produce a mural that evolves with time, an ever-changing tapestry that reflects the world around it. Several case studies highlight artists using AI as a co-creator, resulting in exhibitions that challenge our understanding of creativity. In these exhibits, AI becomes a paintbrush in the artist's hand, not a replacement but an enhancement, offering new textures and possibilities that expand the artistic toolkit.

AI also steps into the realm of design, where it streamlines processes that once took hours, if not days. Picture this: you're working on a graphic design project with looming deadlines, and instead of pulling an all-nighter, you let AI handle the grunt work. Tools like Canva integrate AI to assist with layout design and color matching, ensuring your projects look professional without the stress. It's like having a design assistant who never tires, always ready to tweak that color palette or suggest a different font. AI takes care of repetitive tasks, allowing designers to focus on innovation and creativity, rather than getting bogged down by minutiae.

However, as AI strides confidently into the art world, it kicks up a cloud of debate over authorship and originality. If a machine can create a painting, who owns it? Is it the programmer, the user, or perhaps the machine itself? These questions spark discussions on the value of human creativity versus AI output. Some argue that AI-generated art lacks the soul and intent that human artists imbue into their work. Others see AI as a tool that democratizes creativity, offering everyone a chance to create regardless of traditional skills. It's like debating whether a robot chef can ever truly capture the essence of a grandmother's apple pie.

Interactive Element: Art Creation Challenge

Try your hand at creating AI-generated art using platforms like DeepArt or Artbreeder. Choose a favorite photo and transform it with different styles. Reflect on the process: How does it feel to collaborate with AI in creation? What creative decisions did you make along the way? Share your final piece with friends and discuss the experience of blending human intuition with digital intelligence.

So, whether you're an artist looking to collaborate with a digital muse or a designer eager to automate tedious tasks, AI opens up a world of possibilities. It invites you to explore new frontiers, challenge conventions, and redefine what it means to be creative in the 21st century.

Prompt 1: DeepArt Transformation Challenge

> "Upload a photo that holds personal meaning to you — it could be a favorite travel memory, a beloved pet, or even a simple coffee cup on your desk. Then, use DeepArt to apply three radically different art styles (think Van Gogh's swirls, a pop art explosion, or a futuristic digital landscape). As you transform your image, pay attention: Which styles feel closest to your vision? Which surprise you? Reflect on how it feels to see your memory reimagined through an AI artist's 'eyes.'"

Prompt 2: Artbreeder Remix Adventure

> "Start with a random image on Artbreeder or upload your own. Then, using the 'genes' sliders, experiment with blending different features — like turning a serene landscape into a surreal dreamscape, or mixing human portraits with fantastical creatures. How much did you guide the process versus letting happy accidents happen? Reflect on the collaboration: Were you steering the art, or did it feel like AI was a co-creator nudging you toward unexpected beauty?"

Music and AI: Composing with Algorithms

Imagine a studio where the walls hum with the potential of melodies yet to be born, and in the corner sits a computer, not just passively observing but actively creating music. This is the new frontier of music composition, where AI steps into the role of composer, crafting original pieces and soundtracks that captivate audiences across the globe. Tools like Amper Music and AIVA are at the forefront, offering platforms where technology and creativity dance in harmony. Amper Music allows users to generate unique music tracks with minimal input, making it a favorite among content creators who need soundtracks that align perfectly with their vision. Meanwhile, AIVA focuses on composing soundtracks for ads, games, and movies, demonstrating AI's ability to create music that resonates with the emotional tone of visual media.

For music producers, AI is like the best technician you could ask for, one that never gets tired or distracted. It enhances sound quality and production efficiency, using AI-driven audio mixing and mastering tools to polish tracks to perfection. LANDR's AI-powered mastering platform is a prime example, offering automated audio mastering that ensures your music sounds great across all devices and platforms. This tool analyzes your track and applies the best settings for optimal sound quality, saving musicians time and ensuring a professional finish. It's like having a seasoned producer in your back pocket, always ready to tweak the bass or polish the vocals to make sure everything hits just right.

Collaborations between AI and musicians are pushing the boundaries of what we thought possible, giving rise to new genres and styles. Albums featuring AI-composed tracks are not just novelties but serious contenders in the music world. Musicians are using AI as a tool for live improvisation, creating performances that blend human spontaneity with machine precision. Imagine a jazz ensemble where the saxophonist is a computer, responding to the nuances of the human band members, adding layers of complexity and surprise to the performance. These collaborations highlight AI's potential not just as a tool but as a partner in the creative process, opening doors to musical possibilities that were previously unimaginable.

However, the marriage of AI and music isn't without its challenges. Licensing and copyright issues loom large, as the question of ownership in AI-generated music remains a thorny issue. If a machine composes a symphony, who gets the credit? These questions are still being debated, as the music industry grapples with how to adapt existing frameworks to accommodate AI's growing role. Despite these challenges, AI's potential to democratize music production is undeniable. With these tools, anyone with a creative spark can produce high-quality music without needing a professional studio or a team of sound engineers. It's like having a full orchestra at your fingertips, ready to bring your musical visions to life.

Interactive Element: AI Music Creation

Experiment with Amper Music or AIVA to create your own original music track. Consider how the process compares to traditional music composition. Reflect on the creative decisions you made and how AI influenced your choices. Share your track with friends and discuss how AI might change the future of music production and consumption.

AI in music is a symphony of innovation, challenging us to rethink what music creation means in a digital age. It's not about replacing musicians but augmenting their creativity, offering new tools and ideas that inspire the next wave of musical innovation. Whether you're a musician, a producer, or just someone who loves a good tune, AI invites you to explore, experiment, and play in the vast soundscape it opens up.

Prompt 1: Amper Music Composition Challenge

> "Use Amper Music to compose a short original track. Start by choosing a mood (e.g., uplifting, mysterious, cinematic) and let Amper generate a base. Then, tweak the instrumentation, tempo, and key to make it your own. As you shape the music, reflect: Were you directing the composition, or was the AI leading you toward unexpected choices? How did it feel to fine-tune a track without ever picking up an instrument?"

Prompt 2: AIVA Storytelling Through Sound

> "Create a piece of music on AIVA that tells a story — imagine you're scoring a short film scene, a quiet moment in nature, or a heroic adventure. Customize the style, emotion, and arrangement settings to match the story you envision. Afterward, reflect: How did the AI enhance or challenge your creative instincts? Did it offer sounds or ideas you wouldn't have considered on your own?"

AI in Writing: Enhancing Creativity

Picture yourself staring at a blank page, the cursor blinking in silent judgment, waiting for inspiration to strike. This is where AI swoops in like a caped hero ready to save the day. With tools like Grammarly and Sudowrite, AI becomes your creative sidekick, offering suggestions that spark new ideas and refine your prose. Grammarly goes beyond basic spellcheck, providing insights into tone and style, ensuring your words convey the right message. Meanwhile, Sudowrite acts like a brainstorming buddy, generating plot outlines and character developments that breathe life into your stories. It's like having a writer's room in your pocket, with AI as your ever-ready collaborator.

Beyond the creative spark, AI takes on the nitty-gritty of editing and proofreading, bringing a level of precision that makes even the most meticulous editor nod in approval. Tools like ProWritingAid streamline the editorial process, offering advanced style and grammar correction features that polish your work to a shine. Imagine having a virtual editor that never tires, catching those pesky typos and awkward phrases that slip through the cracks. AI becomes the diligent proofreader, ensuring your manuscript is ready to dazzle agents and publishers alike without the hassle of manual revisions.

In the realm of digital media, AI is the unsung hero behind the scenes, generating and curating content at a pace that would leave human writers breathless. News articles and blog posts are crafted by algorithms capable of churning out coherent, relevant content with remarkable speed. AI analyzes trends, compiles data, and structures information into articles that keep audiences informed and engaged. It's like having a newsroom full of tireless reporters, all focused on delivering the latest scoop without the need for coffee breaks or late-night pizza runs.

However, as AI becomes more adept at writing, it stirs up debates over originality and credibility. If an algorithm can pen an article, where does that leave human writers? Concerns about plagiarism and authenticity loom large, as the line between human and machine-generated content blurs. This raises questions about the role of AI in academia, where originality is paramount. Some fear that reliance on AI could undermine the credibility of authors, as the distinction between human creativity and machine efficiency becomes increasingly challenging to discern. It's like having a ghostwriter who never takes credit, crafting pieces that may lack the soul and nuance of a human touch.

Reflection Section: Exploring AI's Impact on Your Writing

Consider how AI tools have influenced your writing process. Have they sparked new ideas or streamlined your workflow? Reflect on the benefits and challenges of integrating AI into your creative practices. How do you balance AI's efficiency with maintaining your authentic voice?

As writers navigate these evolving landscapes, AI serves as both a powerful ally and a source of existential questions. It challenges traditional notions of authorship, pushing us to redefine creativity in an age where machines can write, edit, and curate as proficiently as humans. While AI offers tools that enhance productivity and innovation, it also invites us to ponder the essence of creative expression. It's a call to embrace the possibilities while safeguarding the unique spark that makes each writer's voice distinct.

The Role of AI in Film and Media Production

Imagine sitting in a cozy cinema, popcorn in hand, as the lights dim and a world of spectacle unfolds on screen. Behind the scenes, AI has quietly revolutionized film and media production, reshaping everything from scriptwriting to visual effects. In the realm of storytelling, AI tools have become invaluable collaborators, generating plot ideas and character arcs that bring fresh perspectives to the narrative process. ScriptBook, for instance, analyzes scripts to predict their potential success, offering insights that guide writers in crafting compelling stories. It's like having a wise script doctor on standby, ready to diagnose any narrative hiccups and prescribe the perfect plot twist.

Visual effects, the dazzling spectacles that transport us to otherworldly realms, are another area where AI flexes its creative muscles. AI enhances CGI and animation, creating visuals that are not only more realistic but also more efficiently produced. Real-time rendering powered by AI allows filmmakers to see the results of their digital wizardry instantly, reducing the time spent waiting for renders to complete. This rapid feedback loop lets directors and animators iterate on their ideas quickly, ensuring the final product is as polished as a diamond. With AI, even the most intricate scenes come to life with breathtaking detail, capturing audiences' imaginations with seamless artistry.

Post-production, often a painstaking process of piecing together the cinematic puzzle, benefits immensely from AI's precision and speed. Tools like Adobe's Sensei streamline editing by automating scene detection and color grading, leaving editors free to focus on the creative nuances of storytelling. AI-driven video editing software analyzes footage, identifies key scenes, and suggests edits, transforming raw footage into a coherent narrative with impressive efficiency. It's akin to having a meticulous assistant who never misses a beat, always ready to fine-tune the film until every frame sings in harmony.

Once the film is complete, AI continues to play a pivotal role in content delivery, ensuring audiences receive a tailored viewing experience. Streaming platforms use AI algorithms to recommend content based on user preferences, much like a savvy friend who always knows what show you'll love next. These algorithms analyze viewing habits, genres, and even the time of day you watch, curating a lineup that's as personal as a mixtape from a close friend. Beyond recommendations, AI customizes the viewing experience itself, adjusting aspects like subtitles and audio tracks to match your preferences, providing a seamless and engaging encounter.

In this digital age, the magic of movies and media is brought to life by the tireless efforts of AI, working behind the curtain to create spectacles that captivate and inspire. AI tools not only enhance the creative process but also streamline production, making it possible for creators to push the boundaries of what's possible on screen. Whether it's generating scripts, crafting stunning visuals, or delivering personalized content, AI stands as a silent partner, empowering filmmakers and media producers to tell their stories with unprecedented creativity and precision.

AI in Gaming: Designing Immersive Experiences

Imagine stepping into a video game world where the landscapes stretch on endlessly, and every character you meet has a mind of its own, reacting to your every move with uncanny realism. This isn't a wild fantasy—it's the revolution AI is bringing to game development and design. AI is shaking up how game worlds are created, using procedural content generation to craft vast, dynamic environments that keep players engaged. Think of it like a digital architect tirelessly building intricate landscapes that evolve based on player actions. These environments aren't just static backdrops; they're living, breathing worlds that adapt and change, creating a truly immersive experience.

Non-player characters, or NPCs, are no longer the static, predictable entities they once were. Thanks to AI, they now have behaviors and decision-making abilities that rival those of human players. Imagine encountering an NPC that learns from your interactions, adapting its strategies and responses to keep you on your toes. It's like playing chess with an opponent who gets smarter with each move, challenging you to think beyond the obvious. This kind of dynamic AI-driven NPC behavior adds layers of complexity and excitement to games, making every encounter unique and unpredictable.

But AI's role doesn't stop at making game worlds more expansive and NPCs more lifelike. It's also enhancing player experiences through adaptive gameplay, where difficulty levels and challenges adjust according to each player's skill and preferences. Picture a game that senses when you're breezing through levels and ramps up the challenge, or conversely, offers assistance when you're stuck. This personalization ensures that every player has a tailored experience, maintaining the perfect balance between challenge and enjoyment. It's like having a personal trainer in your virtual world, constantly adjusting the workout to keep you motivated and engaged.

In the realm of game testing and quality assurance, AI is proving to be an invaluable ally. Traditionally, game testing involved countless hours of manual playthroughs, searching for bugs and glitches. Now, AI tools automate this process, running testing scripts that simulate gameplay scenarios to uncover issues that might have slipped through human testers' fingers. It's like having a vigilant watchdog that tirelessly sniffs out problems before they become game-breaking bugs. AI can even detect graphical glitches, ensuring that every pixel is in its rightful place, and the visual experience is as seamless as possible.

As we peer into the future, the integration of AI with virtual reality promises to push the boundaries of immersive gaming even further. Imagine donning a VR headset and entering a world where AI not only shapes the environment but also personalizes the storytelling experience. AI's role in creating interactive narratives means that the story evolves based on your choices, leading to multiple endings and unique plot twists. It's like being the director of your own blockbuster movie, where every decision you make influences the outcome, keeping you at the edge of your seat.

In the world of gaming, AI is not just a tool—it's a game-changer, opening doors to experiences that were once the stuff of dreams. By crafting dynamic worlds, enhancing character interactions, personalizing gameplay, and streamlining testing, AI is redefining what it means to be a gamer. The future is bright, with endless possibilities for innovation and creativity, and whether you're a developer, a player, or just someone with a passion for games, AI invites you to explore and shape the next generation of interactive entertainment.

And so, as you close this chapter and look forward to the next, consider how AI is not just reshaping industries but transforming the very fabric of our digital lives, promising to enrich the way we experience the world around us.

Chapter Ten

Staying Ahead

Embracing AI's Future Developments

Your coffee's brewing, your toaster's crunching data on your ideal shade of toast, and somewhere in the background, quantum physics is quietly stirring. No, it's not the setup for a futuristic sitcom—it's a sneak peek at the kind of world we're rapidly heading toward.

Artificial intelligence is no longer confined to research labs or sci-fi novels. It's elbowing its way into our kitchens, offices, and daily routines, turning ordinary moments into something downright astonishing. As we stand at the edge of this next AI frontier, the question isn't if the future is coming—it's how fast, how weird, and how wonderfully transformative it's going to be.

Let's kick things off with quantum computing, a concept that sounds like it belongs in a sci-fi novel. But trust me, it's very real and could revolutionize AI capabilities. While our current computers are like diligent clerks working through calculations step-by-step, quantum computers are more like super-powered geniuses multitasking with ease. They use qubits, which can exist in multiple states simultaneously, allowing them to perform calculations at lightning speed. This means that tasks that would take traditional computers eons could be resolved in mere seconds. Imagine an AI with the ability to process massive datasets and predict trends with unprecedented accuracy. Companies like Google, IBM, and D-Wave are leading the charge, each racing to unlock the potential of quantum AI. The possibilities are endless—from transforming drug discovery to optimizing supply chains, the impact on industries could be monumental.

Now, let's shift gears and talk about AI hardware. If quantum computing is the brainy genius, then AI-specific hardware is the bodybuilder flexing its muscles. Recent advancements in AI hardware are paving the way for more efficient and powerful AI systems. Neuromorphic computing, inspired by the workings of the human brain, is one such innovation. These brain-inspired processors aim to mimic the brain's ability to process information, potentially leading to smarter and more energy-efficient AI systems. Think of it as upgrading from a trusty old sedan to a sleek, high-performance sports car. Meanwhile, AI accelerators like Tensor Processing Units (TPUs) are designed to handle the heavy lifting of AI tasks. They're like the workhorses of the tech world, delivering the raw power needed to train complex models faster and more effectively.

As we edge closer to ubiquitous computing, AI is finding its way to the edge—literally. Edge computing is bringing intelligence closer to where data is generated, enabling real-time processing in devices like smartphones and IoT gadgets. Imagine an autonomous vehicle that processes data on-the-fly, reacting to obstacles in a split second without relying on a distant data center. Edge computing makes this possible by reducing latency and improving responsiveness. It's like having a personal assistant who's not only quick on their feet but always right by your side. Smart cities, too, stand to benefit, with AI-driven systems managing everything from traffic flow to energy usage, creating environments that are as efficient as they are futuristic.

And then there's 5G, the tech world's latest buzzword. When paired with AI, 5G technology is set to transform the landscape of connectivity. With its blazingly fast speeds and low latency, 5G enhances the capabilities of AI-driven devices, enabling seamless communication and data transfer.

Picture remote surgeries performed with precision, aided by real-time data exchange, or augmented reality experiences that blend seamlessly with the real world. The synergy between AI and 5G unlocks new applications, turning science fiction into everyday reality. It's like upgrading from dial-up to fiber optic overnight—suddenly, the world is faster, more connected, and bursting with potential.

Visual Element: AI and 5G Infographic

To visualize the impact of AI and 5G, here's a quick infographic highlighting key benefits:

- Enhanced connectivity for AI-driven devices
- Real-time data exchange enabling new applications
- Improved precision and efficiency in fields like healthcare and augmented reality

As we embrace these emerging technologies, it's clear that the future of AI is bright and brimming with possibilities. From quantum computing's transformative potential to the power of edge computing and 5G connectivity, we stand on the cusp of a new era. So, keep your eyes peeled and your mind open because the next big thing in AI is just around the corner.

Preparing for AI's Impact on Society and Economy

Imagine waking up and finding yourself in a world where your favorite barista is now a robot, and your local delivery driver is a self-driving van. As AI continues to weave itself into the very fabric of our daily lives, its impact on the workforce is becoming more apparent. AI is poised to automate repetitive tasks that have long been the bread and butter of many jobs.

While this may sound like a dream come true for those who dread the monotony of data entry or assembly lines, it does raise questions about job displacement. On the flip side, AI is also birthing new, exciting opportunities. Entire industries are emerging that focus on developing and managing AI technologies. This means that while some traditional roles may fade away, new ones are cropping up faster than you can say "machine learning." For those willing to pivot, the AI revolution could mean stepping into roles that didn't even exist a few years ago. AI specialists, data scientists, and automation consultants are just a few of the new careers that are gaining traction.

The economic implications of this shift are worth pondering. AI's adoption across various industries is set to turbocharge productivity, leading to significant growth, particularly in tech-heavy sectors. Businesses are finding that they can achieve more with less, thanks to AI's efficiency. This newfound productivity could potentially elevate entire economies, reshaping the global economic landscape. Imagine countries traditionally on the tech periphery suddenly becoming powerhouses due to strategic AI investments. As AI technologies mature, they could very well shift the global power dynamics, with nations competing to become the next AI leader. This shift could lead to more collaborative international relations, as countries work together to harness AI's potential. However, it could also spark competition, as each nation races to outdo the other in AI advancements.

But before we get too carried away with visions of a utopian AI future, it's important to address the social equity challenges that AI presents. While AI has the potential to level the playing field in areas like healthcare and education, it also risks exacerbating existing inequalities. On one hand, AI could democratize access to quality healthcare by enabling telemedicine and personalized treatment plans that were previously out of reach for many. In education, AI could tailor learning experiences to individual students' needs, making personalized education a reality rather than a privilege. On the other hand, there is a risk that those without access to AI technologies could be left further behind. The digital divide could widen if AI is not implemented inclusively. This makes it crucial for policymakers to ensure that AI technologies are accessible to all, and not just a select few.

This brings us to the importance of policy and regulation in AI integration. Governments have a key role to play in setting the stage for responsible AI adoption. Initiatives that fund AI research and development are vital, as they ensure that advancements in AI are aligned with societal needs. International collaborations on AI governance can help establish standards and frameworks that promote ethical AI use worldwide. Picture a global AI council, where countries come together to share knowledge, set ethical guidelines, and collaborate on AI projects that benefit humanity. Such efforts could help mitigate the risks associated with AI, ensuring that its benefits are distributed equitably.

Regulatory frameworks can also address potential ethical concerns that arise from AI's integration into society. Privacy, for instance, becomes a hot topic when AI systems start collecting and analyzing vast amounts of personal data. Ensuring robust data protection laws and advocating for transparency in AI operations are critical steps in maintaining public trust. Policymakers must also consider the implications of AI on employment. They need to develop programs that support workforce transitions and reskilling initiatives, enabling individuals to adapt to the changing job market. By fostering an environment where AI can thrive responsibly, governments can maximize the benefits of AI while minimizing its potential downsides.

In this rapidly evolving landscape, staying informed and adaptable is key. Keeping an eye on AI trends and developments will enable you to anticipate changes and position yourself advantageously. Whether it's exploring new career paths, advocating for inclusive AI policies, or simply understanding the societal shifts that AI brings, staying engaged will ensure that you can navigate the AI era with confidence. Remember, the future may be unpredictable, but it's also brimming with opportunities for those ready to embrace change.

Building Resilience: Adapting to AI-Driven Change

Alright, folks, let's talk about rolling with the punches in this ever-evolving AI landscape. The key to thriving in a world increasingly dominated by AI is cultivating an adaptive mindset. Think of it as being a lifelong learner, always ready to pick up a new skill or two. Embrace change with open arms and a curious mind. Because, let's face it, AI isn't going anywhere but forward. Skill agility is your best friend here. Staying informed about the latest AI advancements is like keeping up with the latest episodes of your favorite show—you wouldn't want to miss out on the plot twists, right? Subscribe to tech blogs, attend webinars, and join AI forums where the latest buzz gets dissected.

Now, let's switch gears to organizations. Imagine trying to steer a cruise ship through a narrow canal—that's what managing transitions to AI-driven operations can feel like. But fear not, because with the right change management frameworks, you can navigate these waters smoothly. Successful integration of AI into business operations is not just about installing new software; it requires a cultural shift. Companies need to foster an environment that encourages experimentation and adaptation. Take some inspiration from businesses that have nailed this transition, like those in retail that have seamlessly integrated AI for inventory and customer insights. Such success stories highlight the importance of preparing your team for change, providing training, and ensuring that everyone is on board with the new direction.

Let's chat about the backbone of any successful AI initiative—technological infrastructure. Just like you wouldn't build a house without a strong foundation, you can't support AI projects without robust infrastructure. Investment in cloud computing and data storage solutions is crucial for handling the massive amounts of data AI systems require. Think of cloud storage as your digital attic, spacious and organized, ready to store all the files and data you throw its way. But don't forget about cybersecurity—it's like the lock on your front door, keeping your AI systems safe from unwanted intruders. With cyber threats on the rise, enhancing cybersecurity measures is not just a good idea; it's a necessity.

While we're on the subject of teamwork, fostering collaboration and innovation is vital for driving AI forward. No one can do it alone, and the best ideas often come from bouncing thoughts off others. Partnerships between academia, industry, and government can create a fertile ground for AI innovation. It's like bringing together a team of superheroes, each with unique powers, to tackle challenges that one simply can't handle alone. Collaborative platforms for AI research and development are springing up everywhere, offering opportunities for cross-pollination of ideas. These platforms serve as bridges, connecting different sectors and fostering an environment where innovation can thrive.

So there you have it, a sneak peek into building resilience in an AI-driven world. Embracing change, investing in the right infrastructure, and fostering collaboration are just a few ways to stay ahead. As AI continues to evolve, these strategies will ensure you're not just surviving but thriving in this exciting new era.

Chapter Eleven

Conclusion

Well, folks, we've come to the end of our AI adventure, and what a wild ride it's been! Together, we've navigated the twists and turns of artificial intelligence, from understanding the basics to mastering advanced applications. We've laughed, we've learned, and we've even tackled some pretty heavy topics along the way.

Let's take a quick stroll down memory lane, shall we? We started by demystifying AI, breaking down its components and exploring its real-world applications. You discovered how AI is already woven into the fabric of our daily lives, from the smartphones in our pockets to the algorithms that curate our social media feeds.

Next, we dove into the nitty-gritty of AI tools, equipping you with the skills to harness their power. You learned how to choose the right tools for your needs, got hands-on with chatbots, and even explored advanced techniques in prompt engineering. By the end of that chapter, you were ready to take on AI like a pro.

But we didn't stop there. We explored how AI can revolutionize your productivity, streamline your business operations, and even tap into your creative potential. You discovered AI's role in marketing, customer service, and financial forecasting. We even took a wild ride through the creative industries, seeing how AI is transforming art, music, and gaming.

Of course, we couldn't ignore the ethical considerations that come with AI's rapid advancement. We grappled with questions of bias, privacy, and the need for responsible AI practices. You learned how to navigate these challenges and advocate for ethical AI integration.

Throughout this journey, my goal has been to empower you to embrace AI with confidence and creativity. I want you to walk away from this book feeling like you can tackle any AI challenge that comes your way. Whether you're looking to boost your career, start a side hustle, or simply stay ahead of the curve, you now have the tools and knowledge to succeed.

But the learning doesn't stop here. AI is a field that's constantly evolving, with new breakthroughs happening every day. To stay on top of your game, keep exploring, keep experimenting, and keep pushing the boundaries of what's possible. Attend workshops, join online communities, and never stop asking questions.

As we look to the future, one thing is clear: AI is here to stay, and its potential is limitless. From quantum computing to 5G connectivity, the next wave of AI innovations is just around the corner. By staying engaged and adaptable, you'll be ready to ride that wave to exciting new heights.

So, my friend, as you close this book and step out into the world, remember that you have the power to shape the future of AI. Whether you're developing cutting-edge applications or simply advocating for responsible AI practices, your voice matters. Embrace the challenges, celebrate the victories, and never stop learning.

In the words of my favorite AI philosopher (okay, I may have made that up), "The true measure of intelligence is not knowledge, but imagination." So go forth, imagine boldly, and let AI be your partner in turning those dreams into reality.

It's been an honor to be your guide on this journey, and I can't wait to see where your AI adventures take you next. Remember, the future is yours to create – so make it a good one!

References

- *AI History - Artificial Intelligence (AI) - Library Guides at Ohio ...* https://libguides.library.ohio.edu/AI/history
- *Understanding the different types of artificial intelligence* https://www.ibm.com/think/topics/artificial-intelligence-types
- *Deep Learning vs. Machine Learning: A Beginner's Guide* https://www.coursera.org/articles/ai-vs-deep-learning-vs-machine-learning-beginners-guide#:~:text=Machine%20learning%20and%20deep%20learning,process%20of%20the%20human%20brain.
- *8 Natural Language Processing (NLP) Examples* https://www.tableau.com/learn/articles/natural-language-processing-examples
- *20 AI Tools To Supercharge Your Business And Productivity* https://www.forbes.com/sites/jodiecook/2023/09/11/20-ai-tools-to-supercharge-your-business-and-productivity/
- *Google Cloud AI Platform vs IBM Watson Machine Learning* https://www.peerspot.com/products/comparisons/google-cloud-ai-platform_vs_ibm-watson-machine-learning#:~:text=Features%3A%20IBM%20Watson%20Machine%20Learning,APIs%2C%20and%20seamless%20service%20integration.

- *50 ChatGPT Use Cases with Real-Life Examples in Feb/2025* https://research.aimultiple.com/chatgpt-use-cases/
- *A Guide to Using and Troubleshooting AI Tools | Jitender K.* https://www.linkedin.com/pulse/conquering-ai-frontier-guide-using-troubleshooting-tools-rohtaki-hi90c
- *We Tested 50+ AI Productivity Tools. Here Are The 16 Best ...* https://www.usemotion.com/blog/ai-productivity-tools
- *Partnering with AI: the case of digital productivity assistants* https://pmc.ncbi.nlm.nih.gov/articles/PMC11459757/
- *Maximize Your Time Management: 4 Tips Using AI Tools ...* https://www.forbes.com/sites/lucianapaulise/2024/10/22/maximize-your-time-management-4-tips-using-ai-tools-and-automation/
- *Best 11 AI Solutions for Data Management and Security* https://www.velotix.ai/resources/blog/best-ai-solutions-for-data-management-and-security/
- *AI And Personalization In Marketing* https://www.forbes.com/councils/forbescommunicationscouncil/2024/01/05/ai-and-personalization-in-marketing/
- *The best AI chatbots in 2023 - Section School* https://www.sectionschool.com/blog/the-best-ai-chatbots-in-2023
- *Harnessing AI's Potential To Revolutionize Financial ...* https://www.forbes.com/councils/forbesbusinesscouncil/2023/11/20/harnessing-ais-potential-to-revolutionize-financial-forecasting/
- *Artificial intelligence in supply chain and operations*

... https://www.tandfonline.com/doi/full/10.1080/00207543.2023.2232050

- *How To Write Amazing Generative AI Prompts* https://www.forbes.com/sites/bernardmarr/2024/09/25/how-to-write-amazing-generative-ai-prompts/
- *Prompt Engineering Guide* https://www.promptingguide.ai/
- *Beyond the Hype: Real-World Examples of How Prompts* ... https://medium.com/@simrwaraich/beyond-the-hype-real-world-examples-of-how-prompts-make-ai-work-1fef905ecb40
- *Unlocking Business Growth with Prompt Engineering ...* https://pcg.io/insights/prompt-engineering-business-growth/#:~:text=By%20automating%20and%20optimizing%20key,can%20deliver%20measurable%20business%20value.
- *Monetizing AI in 2024: Emerging Trends and Opportunities* ... https://medium.com/tech-encyclopedia/monetizing-ai-in-2024-emerging-trends-and-opportunities-to-watch-532598eaf171
- *10 AI Skills To Boost Your Salary Up To 47% In 2025* https://www.forbes.com/sites/rachelwells/2025/02/13/10-ai-skills-to-boost-your-salary-up-to-47-in-2025/
- *Case Studies: Successful Entrepreneurs Who Utilized GPT ...* https://medium.com/@DigitalQuill.ai/case-studies-successful-entrepreneurs-who-utilized-gpt-ai-chatgpt-startup-a96fb6a47ed5
- *How to Use AI for Competitive Market Analysis* https://www.talonic.com/blog/how-to-use-ai-for-competitive-market-analysis
- *How are Google and Microsoft Handling Ethical Issues in AI?* https://www.bramblehub.co.uk/news/article/microsoft-go

ogle-responsible-ai

- *Biased Technology: The Automated Discrimination* ... https://www.aclu-mn.org/en/news/biased-technology-automated-discrimination-facial-recognition

- *The Impact of the GDPR on Artificial Intelligence* https://securiti.ai/impact-of-the-gdpr-on-artificial-intelligence/

- *AI Ethics in Business: Artificial Intelligence Policies and ...* https://www.aiguardianapp.com/post/ethics-of-ai-in-business

- *Generative AI and the future of work in America* https://www.mckinsey.com/mgi/our-research/generative-ai-and-the-future-of-work-in-america

- *11 Essential Skills for a Job in Artificial Intelligence* https://ep.jhu.edu/news/11-essential-skills-for-a-job-in-artificial-intelligence/

- *10 Top Artificial Intelligence Certifications And Courses For* ... https://www.techtarget.com/whatis/feature/10-top-artificial-intelligence-certifications-and-courses

- *Top AI/ML Events You Should Not Miss in 2023* https://digitalgiraffes.com/top-ai-ml-events-you-should-not-miss/

- *DeepArt vs. Artbreeder: Which AI Transforms Images into* ... https://www.geeksforgeeks.org/deepart-vs-artbreeder-ai-transforms-images-into-artwork/

- *GAN computers generate arts? A survey on visual ...* https://www.sciencedirect.com/science/article/abs/pii/S0141938222000658

- *AI for Music Production: 10 Tools to Produce Like a Pro* https://dittomusic.com/en/blog/ai-for-music-production-t

ools-for-musicians

- *Impact of AI on Film Post-Production Processes - Vitrina* https://vitrina.ai/blog/impact-of-ai-on-film-post-production-processes/
- *How Quantum AI Will Reshape Our World* https://www.forbes.com/sites/bernardmarr/2024/10/08/the-next-breakthrough-in-artificial-intelligence-how-quantum-ai-will-reshape-our-world/
- *What happened at IBM Research's 2023 AI Hardware Forum* https://research.ibm.com/blog/ai-hardware-forum-2023
- *Opportunities and challenges of 5G network technology ...* https://pmc.ncbi.nlm.nih.gov/articles/PMC10651640/
- *Generative AI and the future of work in America* https://www.mckinsey.com/mgi/our-research/generative-ai-and-the-future-of-work-in-america

About the Author

Roger Best

MEET ROGER BEST

Roger Best is a seasoned entrepreneur, author, and coach with over four decades of experience building and leading successful businesses across multiple industries. From his first venture at the age of eight, digging and selling worms to a local bait shop, Roger has been driven by a relentless curiosity, a sharp entrepreneurial mind, and a desire to create meaningful success, not just financial, but personal and lasting.

After years of pushing through 60+ hour workweeks, skipped vacations, and the ever-elusive promise of "one day," Roger experienced what so many entrepreneurs do: burnout. But rather than accept it as the cost of doing business, he chose a different path, one that led to balance, freedom, and the ability to build a life he didn't need to escape from.

Today, Roger lives and works in Puerto Rico with his wife of more than 46 years, where he runs multiple businesses, writes, coaches, and enjoys the sunshine and laid-back island vibes. His books have helped countless entrepreneurs trade exhaustion for intentional living, and reclaim the freedom they originally set out to find.

Roger's mission is simple but powerful: to help entrepreneurs escape the grind, rediscover their purpose, and design businesses that support the life they actually want. Through personal storytelling, practical strategies, and a healthy dose of humor, his writing resonates with overworked professionals who know there must be a better way and are ready to find it.

Whether through his books, one-on-one coaching, webinars, or mastermind groups, Roger is your guide to building a business that works for you, not the other way around. You'll find me online at RogerGBest.com.

Also by Roger Best

Explore a collection of inspiring and practical reads designed to empower entrepreneurs and business owners. From reclaiming your time to building a business that truly serves you, these books offer actionable insights to help you create a successful and meaningful life and business.

From Hamster Wheel to Hammock: A Guide to Taking Your Day Back*(Available on Amazon – ebook, paperback and Audible)*

Escape the relentless grind and rediscover the freedom you dreamed of when you started your business. This practical and empowering guide offers actionable strategies to help you break free from the hamster wheel, reclaim your time, and build a balanced, fulfilling life. Whether you're overwhelmed by long hours or stuck in the cycle of constant hustle, this book will show you the path to a life of purpose and peace.

Personal Growth for Entrepreneurs: Your Time, Your Way*(Available on Amazon – ebook and paperback)*

Take a deeper dive into aligning your business and life with your values and passions. This book focuses on developing habits, mindsets, and systems that not only support your professional goals but also nurture your personal growth. Learn how to create meaningful success that feels true to who you are while finding fulfillment in every step of the journey.

The Liberated Entrepreneur: Building a Business That Works for You (Available on Amazon – ebook and paperback)

Running a business doesn't have to mean sacrificing your life. In this transformative guide, you'll learn how to build a business that aligns with your values, supports your dreams, and works *for* you—not the other way around. Discover strategies for streamlining operations, empowering your team, and creating a sustainable work-life balance. This book is for entrepreneurs ready to escape burnout and reclaim the joy and freedom they set out to achieve.

Purpose-Driven Entrepreneurship: Reclaim Your Why for Lasting Success *(Available on Amazon – ebook and paperback)*

Discover the power of reconnecting with your purpose and unlock the clarity and passion needed to thrive. This book offers practical exercises, journaling prompts, and reflective tools to help you rediscover your core motivations, make aligned decisions, and build a life and business that truly matter. Perfect as a companion to *From Hamster Wheel to Hammock* and *Personal Growth for Entrepreneurs*, this book is essential for anyone seeking deeper connection, fulfillment, and long-term success.

Protecting Your Business in the Digital Age: A Non-Technical Guide to Cybersecurity for Small Business Owners (Available on Amazon – ebook, paperback, and Audible)

Cybersecurity doesn't have to be overwhelming or overly technical. In this straightforward guide, you'll learn how to protect your business from digital threats, safeguard your financial resources, and ensure your customer data stays secure. Designed specifically for small business owners, this book breaks down the essentials of cybersecurity into practical, easy-to-understand steps that anyone can implement. Whether you're a tech novice or simply want peace of mind, this guide will help you secure your business and its future.

Each of these books offers unique insights into different aspects of entrepreneurial life. Whether you're seeking to break free from burnout, deepen your personal growth, reconnect with your purpose, or protect your business in a rapidly changing digital landscape, these reads are your guide to a more intentional, successful, and resilient journey. Explore them today and take the next step toward the life and business you've always envisioned!

Made in the USA
Monee, IL
09 November 2026

67381878R00098